Everyday Mathematics®

The University of Chicago School Mathematics Project

STUDENT MATH JOURNAL

VOLUME 2

Mc Graw Hill Education

The University of Chicago School Mathematics Project

Max Bell, Director, *Everyday Mathematics* First Edition; James McBride, Director, *Everyday Mathematics* Second Edition; Andy Isaacs, Director, *Everyday Mathematics* Third, CCSS, and Fourth Editions; Amy Dillard, Associate Director, *Everyday Mathematics* Third Edition; Rachel Malpass McCall, Associate Director, *Everyday Mathematics* CCSS and Fourth Editions; Mary Ellen Dairyko, Associate Director, *Everyday Mathematics* Fourth Edition

Authors
Max Bell, Jean Bell, John Bretzlauf, Amy Dillard, Robert Hartfield, Andy Isaacs, James McBride, Cheryl G. Moran, Kathleen Pitvorec, Peter Saecker

Fourth Edition Grade 2 Team Leader
Cheryl G. Moran

Writers
Camille Bourisaw, Mary Ellen Dairyko, Gina Garza-Kling, Rebecca Williams Maxcy, Kathryn M. Rich

Open Response Team
Catherine R. Kelso, Leader; Steve Hinds

Differentiation Team
Ava Belisle-Chatterjee, Leader; Jean Marie Capper, Martin Gartzman

Digital Development Team
Carla Agard-Strickland, Leader; John Benson, Gregory Berns-Leone, Juan Camilo Acevedo

Virtual Learning Community
Meg Schleppenbach Bates, Cheryl G. Moran, Margaret Sharkey

Technical Art
Diana Barrie, Senior Artist; Cherry Inthalangsy

UCSMP Editorial
Don Reneau, Senior Editor; Rachel Jacobs, Kristen Pasmore, Luke Whalen

Field Test Coordination
Denise A. Porter

Field Test Teachers
Kristin Collins, Debbie Crowley, Brooke Fordice, Callie Huggins, Luke Larmee, Jaclyn McNamee, Vibha Sanghvi, Brook Triplett

Digital Field Test Teachers
Colleen Girard, Michelle Kutanovski, Gina Cipriani, Retonyar Ringold, Catherine Rollings, Julia Schacht, Christine Molina-Rebecca, Monica Diaz de Leon, Tiffany Barnes, Andrea Bonanno-Lersch, Debra Fields, Kellie Johnson, Elyse D'Andrea, Katie Fielden, Jamie Henry, Jill Parisi, Lauren Wolkhamer, Kenecia Moore, Julie Spaite, Sue White, Damaris Miles, Kelly Fitzgerald

Contributors
William B. Baker, John Benson, Kathleen Clark, Jeanne Mills DiDomenico, James Flanders, Lila K. S. Goldstein, Funda Gönülateş, Allison M. Greer, Lorraine M. Males, John P. Smith III, Patti Satz, Penny Williams

Center for Elementary Mathematics and Science Education Administration
Martin Gartzman, Executive Director; Jose J. Fragoso, Jr., Meri B. Fohran, Regina Littleton, Laurie K. Thrasher

External Reviewers
The *Everyday Mathematics* authors gratefully acknowledge the work of the many scholars and teachers who reviewed plans for this edition. All decisions regarding the content and pedagogy of *Everyday Mathematics* were made by the authors and do not necessarily reflect the views of those listed below.

Elizabeth Babcock, California Academy of Sciences; Arthur J. Baroody, University of Illinois at Urbana-Champaign and University of Denver; Dawn Berk, University of Delaware; Diane J. Briars, Pittsburgh, Pennsylvania; Kathryn B. Chval, University of Missouri-Columbia; Kathleen Cramer, University of Minnesota; Ethan Danahy, Tufts University; Tom de Boor, Grunwald Associates; Louis V. DiBello, University of Illinois at Chicago; Corey Drake, Michigan State University; David Foster, Silicon Valley Mathematics Initiative; Funda Gönülateş, Michigan State University; M. Kathleen Heid, Pennsylvania State University; Natalie Jakucyn, Glenbrook South High School, Glenview, IL; Richard G. Kron, University of Chicago; Richard Lehrer, Vanderbilt University; Susan C. Levine, University of Chicago; Lorraine M. Males, University of Nebraska-Lincoln; Dr. George Mehler, Temple University and Central Bucks School District, Pennsylvania; Kenny Huy Nguyen, North Carolina State University; Mark Oreglia, University of Chicago; Sandra Overcash, Virginia Beach City Public Schools, Virginia; Raedy M. Ping, University of Chicago; Kevin L. Polk, Aveniros LLC; Sarah R. Powell, University of Texas at Austin; Janine T. Remillard, University of Pennsylvania; John P. Smith III, Michigan State University; Mary Kay Stein, University of Pittsburgh; Dale Truding, Arlington Heights District 25, Arlington Heights, Illinois; Judith S. Zawojewski, Illinois Institute of Technology

Note
Too many people have contributed to earlier editions of *Everyday Mathematics* to be listed here. Title and copyright pages for earlier editions can be found at http://everydaymath.uchicago.edu/about/ucsmp-cemse/.

www.everydaymath.com

Send all inquiries to:
McGraw-Hill Education
STEM Learning Solutions Center
8787 Orion Place
Columbus, OH 43240

ISBN: 978-0-02-143086-4
MHID: 0-02-143086-1

Printed in the United States of America.

14 LMN 20

Contents

Unit 5

Unit 6

Unit 7

Unit 8

Unit 9

Activity Sheets

Solving Number Stories

Lesson 5-1

DATE

Solve. You may use the number grid on the inside back cover of your journal to help.

1. Justin brought 30 blueberries for a snack. He ate 10 of them. How many blueberries does he have left?

   Answer: ______ blueberries

2. Davion's string was 48 inches long. He cut off 10 inches. How long is Davion's string now?

   Answer: ______ inches

3. Haily picked up 41 shells on the beach. She picked up 10 more than her sister. How many did her sister pick up?

   Answer: ______ shells

4. Rosa had 108 stamps in her collection. Of these, 78 stamps were from the United States. How many of her stamps were from other countries?

   Answer: ______ stamps

5. The school library ordered 56 new animal books. Children in Ms. Tran's class checked out some of them. The library now has 46 new animal books. How many were checked out?

   Answer: ______ books

Math Boxes

Lesson 5-1

DATE

1. Estimate the length of this line segment:

   About _____ inches

   Use your inch ruler to measure the length of the line segment.

   About _____ inches

   MRB 101–102

2. Write the number word for 98.

   ____

3. Use the digits 6, 1, and 8 to make the smallest number possible.

   _____

   Use the same digits to make the largest number possible.

   _____

   MRB 73

4. Count back by 100s.

   900, _____, _____, _____, _____, _____

5. **Writing/Reasoning** Explain how you made the largest possible number in Problem 3.

   ____

   ____

   ____

   ____

   MRB 73

Math Boxes

DATE

1 Circle the expanded form that shows the larger number.

400 + 20 + 6

400 + 10 + 9

MRB 74

2 Estimate the length of this line segment in centimeters:

About ______ centimeters

Use your centimeter ruler. Measure the length of the line segment.

About ______ centimeters

3 Fill in the bubble that shows the value of the base-10 blocks.

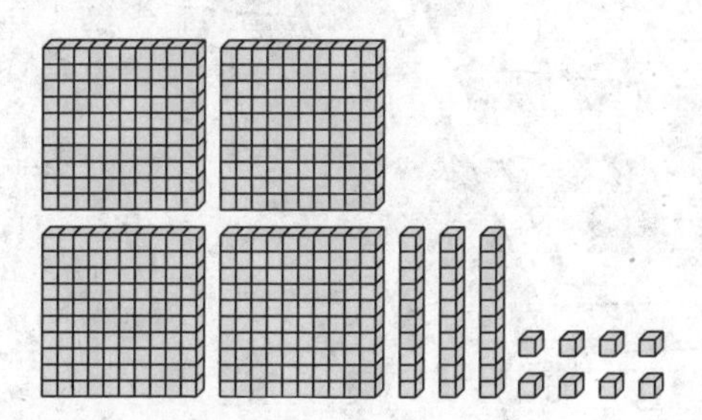

- ◯ 438
- ◯ 483
- ◯ 834
- ◯ 843

4 Write the time.

_____:_____

5 Mark 63 in the correct spot on this number line.

60 ———————————————— 70

Pine School's Fruit and Vegetable Sale

DATE

Buying Fruit and Vegetables at the Sale

Lesson 5-2

DATE

For Problems 1–2 follow these steps:

- Write the name and the cost of one item from journal page 106 that you want to buy.
- Draw the coins you can use to pay for it using Ⓟ, Ⓝ, Ⓓ, and Ⓠ.
- Draw another way to pay using a different combination of coins.

For Problems 3–4 write the names of two or more items you want to buy and how much they cost. Draw coins for the total amount of money you would spend. Then write the total.

I Bought	It Costs	I Paid	I Paid
Example: *Orange*	*18¢*	ⒹⓃ ⓅⓅⓅ	ⓃⓃⓅⓅⓅ ⓅⓅⓅⓅⓅ
1. ______	______		
2. ______	______		
3. ______ ______	______ ______		Total: ______
4. ______ ______ ______	______ ______ ______		Total: ______

Spinning for Money

Lesson 5-2

DATE

Materials
- ☐ *Spinning for Money* Spinner
- ☐ pencil ☐ large paper clip
- ☐ one sheet of paper labeled "Bank"
- ☐ 7 pennies, 5 nickels, 5 dimes, and 4 quarters for each player
- ☐ one $1 bill for the group

Players 2, 3, or 4

Skill Exchange coins and dollar bills

Object of the Game To be first to exchange for a $1 bill

How to Play

1. Each player puts 7 pennies, 5 nickels, 5 dimes, and 4 quarters into the bank. Each group puts one $1 bill into the bank.
2. Players take turns spinning the *Spinning for Money* Spinner and taking the coins shown by the spinner from the bank.
3. Whenever possible, players exchange coins for a single coin or bill of the same value. For example, a player could exchange 5 pennies for a nickel or 2 dimes and 1 nickel for a quarter.
4. The first player to exchange for a $1 bill wins.

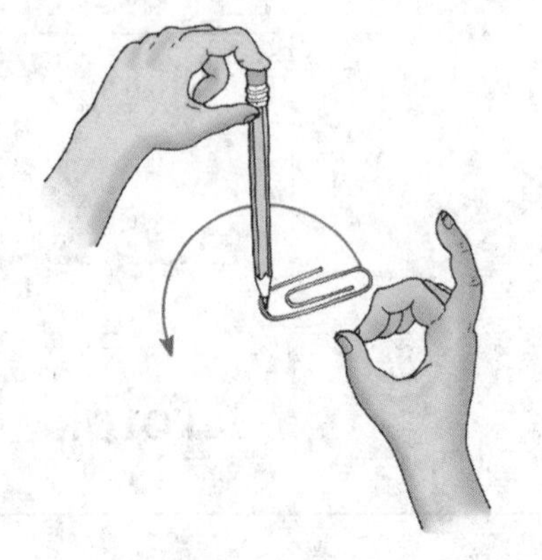

Use a large paper clip and pencil to make a spinner.

Shopping at Pine School's Fruit and Vegetable Sale

Lesson 5-3

DATE

Price per Item					
Pear	13¢	Melon slice	30¢	Lettuce	45¢
Orange	18¢	Apple	12¢	Green pepper	24¢
Banana	9¢	Tomato	20¢	Corn	15¢
Plum	6¢	Onion	7¢	Cabbage	40¢

Complete the table.

I Bought	It Costs	I Paid	I Got in Change
Example: One *Melon slice*	*30¢*	ⓠⓠ	*20¢*
1. One ______	______		______
2. One ______	______		______
3. One ______	______		______
Try This **4.** One ______ and one ______	______ ______		______

Math Boxes

Lesson 5-3

DATE

1. Estimate the length of this line segment:

   About _____ centimeters

   Use your centimeter ruler to measure the length of the line segment.

   About _____ centimeters

   MRB 102, 104

2. Write the number word for 79.

   ____

3. Use the digits 7, 1, and 9 to make the smallest number possible.

   _____

   Use the same digits to make the largest number possible.

   _____

   MRB 73

4. Count back by 10s.

   340, _____, _____, _____, _____, _____

5. **Writing/Reasoning** Suppose you measure the line segment in Problem 1 in inches. Will the number of inches be more than or less than the number of centimeters? Explain.

   ____

   ____

   ____

   MRB 101, 104

Math Boxes

DATE

1 Circle the expanded form that shows the smaller number.

600 + 90 + 5

600 + 80 + 8

MRB 74

2 Estimate the length of this line segment in centimeters:

About ______ centimeters

Use your centimeter ruler. Measure the length of the line segment.

About ______ centimeters

3 Write the number.

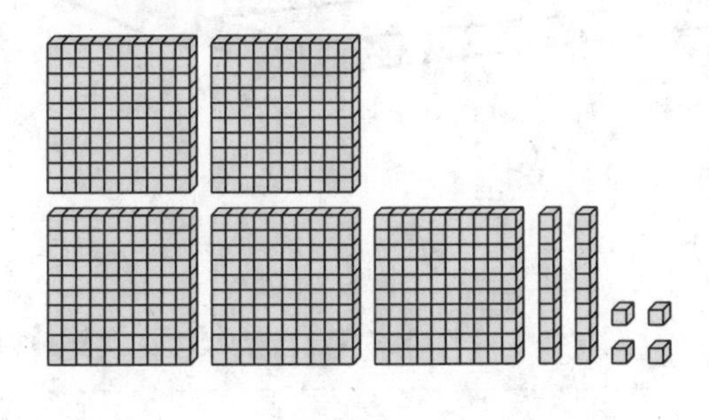

______

MRB 72, 73

4 What time is it? Fill in the bubble next to the best answer.

◯ 6:25

◯ 6:30

◯ 6:35

◯ 7:35

5 Mark 88 in the correct spot on the number line.

80 ———————————————— 90

Buying from a Vending Machine

Lesson 5-4

DATE

1 The exact change light is on.

You Buy	Cost of Item	Draw the coins you put in.
chocolate milk (Chocolate Milk 40¢ PUSH)	40¢	
strawberry yogurt drink (Strawberry Yogurt Drink 70¢ PUSH)	70¢	
____________	______	

Buying from a Vending Machine (continued)

2 The exact change light is off.

You Buy	It Costs	Draw the coins or the $1 bill you put in.	How much change will you get?
2% milk (2% Milk 35¢ PUSH)	35¢	(Q)(Q)	______
chocolate milk (Chocolate Milk 40¢ PUSH)	40¢	$1	______
______	______		______
______	______		______
______	______		______

Try This

3 You want to buy a carton of orange juice with $1. Will you get back enough change to also buy a carton of 2% milk? Explain.

______

______

______

Making *Clock Concentration* Cards

Lesson 5-5

DATE

Materials	☐ 2 copies of *Clock Concentration* Cards, *Math Masters*, p. 130 ☐ scissors

Directions

Make a set of *Clock Concentration* Cards.

1. Cut apart the 5 strips of cards from each copy of *Clock Concentration Cards.* Divide the 10 strips among your group.
2. Draw an hour hand and a minute hand on each clock face.
3. Write the time on the other half of each strip. Check one another's work.
4. Cut each strip in half to make 2 cards.
5. Your group should have 10 cards with a clock face and 10 cards with a time.
6. Write C on the back of each card with a clock face.
7. Write T on the back of each card with a time.
8. Make a mark on each card to show that it belongs to your group.

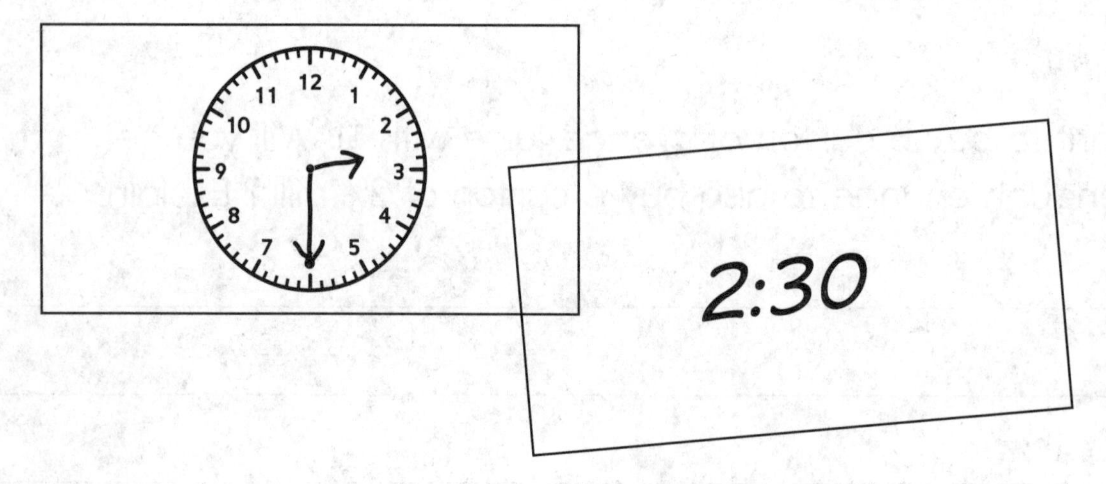

Math Boxes

Lesson 5-5

DATE

Math Boxes

1. In the number 300 there are

   _____ hundreds.

   _____ tens.

   _____ ones.

MRB 73

2. Write <, >, or =.

   549 _____ 595

   378 _____ 308

   956 _____ 856

3. Name something in the classroom that is about 1 foot long.

   ____

4. Draw at least two ways to show 30¢ using Ⓠ, Ⓓ, and Ⓝ.

5. **Writing/Reasoning** For Problem 3, how did you find something that was about 1 foot long?

   ____

   ____

   ____

   ____

Math Boxes

Lesson 5-6

DATE

1. Use centimeters. Estimate the length of this line segment:

____

About ______ centimeters

Use your centimeter ruler. Measure the length of the line segment.

About ______ centimeters

MRB 104

2. Write the time.

______:______

MRB 106–107

3. You buy a banana for 65¢. You pay with 3 quarters. Fill in the bubble that shows how much change you should get.

- ◯ 25¢
- ◯ 15¢
- ◯ 10¢
- ◯ 5¢

4. Write a number with 6 in the hundreds place, 4 in the ones place, and 8 in the tens place.

______

MRB 73

5. You can make 85¢ with

2 Ⓠ and ______ Ⓝ.

You can make 70¢ with

2 Ⓠ and ______ Ⓓ.

MRB 110–111

6. Write the number word for 123.

____

____

Using Open Number Lines to Solve Number Stories

Lesson 5-7

DATE

You are building towers with red and blue blocks. Use an open number line to help you solve each number story. Show your work.

1. You build the first tower with 15 red blocks and 20 blue blocks. How many blocks did you use? ______ blocks

2. You build the second tower with 37 red blocks and 32 blue blocks. How many blocks did you use? ______ blocks

3. You build the third tower with 16 red blocks. You use 39 blocks in all. How many blue blocks did you use? ______ blue blocks

Math Boxes

Lesson 5-7

DATE

1. In the number 500 there are

_____ hundreds.

_____ tens.

_____ ones.

2. Write <, >, or =.

459 _____ 495

207 _____ 179

902 _____ 898

3. Estimate. About how many feet long is your desk from side to side?

About _____ feet

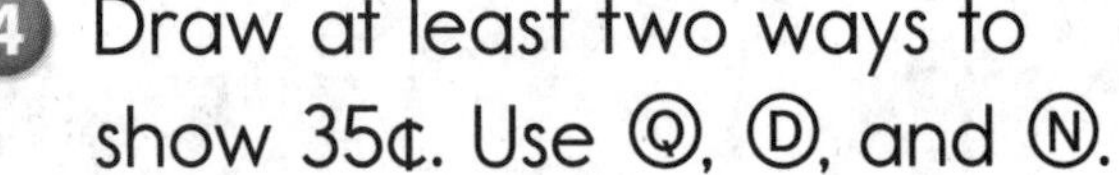

4. Draw at least two ways to show 35¢. Use Ⓠ, Ⓓ, and Ⓝ.

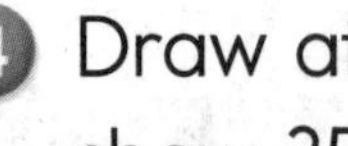

5. **Writing/Reasoning** Explain how you compared the numbers in Problem 2.

____

____

____

____

DATE

Math Boxes
Preview for Unit 6

1 Write each number in expanded form.

34 ____________

66 ____________

52 ____________

81 ____________

MRB 72

2 What number do the base-10 blocks show? ______

Use base-10 shorthand to show the number another way.

3 Hannah scored 10 points. Jen scored 15 points. How many points did they score in all?

Number model:

____

Answer: ______ points

4

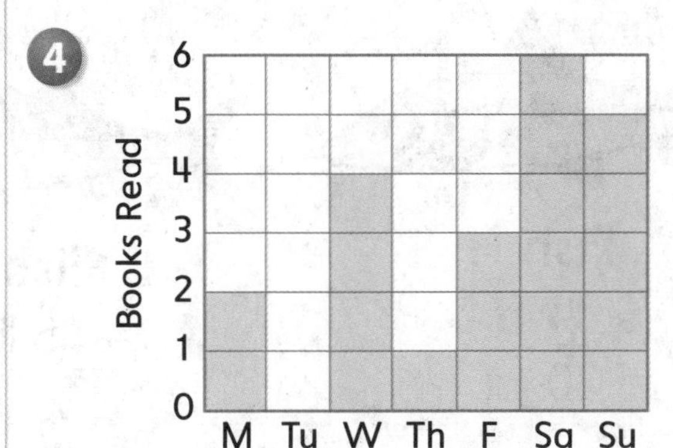

How many more books were read on Saturday than on Thursday? ______ books

5 Chad has 12 pennies. Adam has 7 pennies. How many more pennies does Chad have than Adam? Draw a picture to show what you did to solve this problem.

Answer: ______ pennies

Fish Poster

DATE

Fish A
1 lb
12 in.

Fish B
3 lb
14 in.

Fish C
4 lb
18 in.

Fish D
5 lb
24 in.

Fish E
6 lb
24 in.

Fish F
8 lb
30 in.

Fish G
10 lb
30 in.

Fish H
14 lb
30 in.

Fish I
15 lb
30 in.

Fish J
24 lb
36 in.

Fish K
35 lb
42 in.

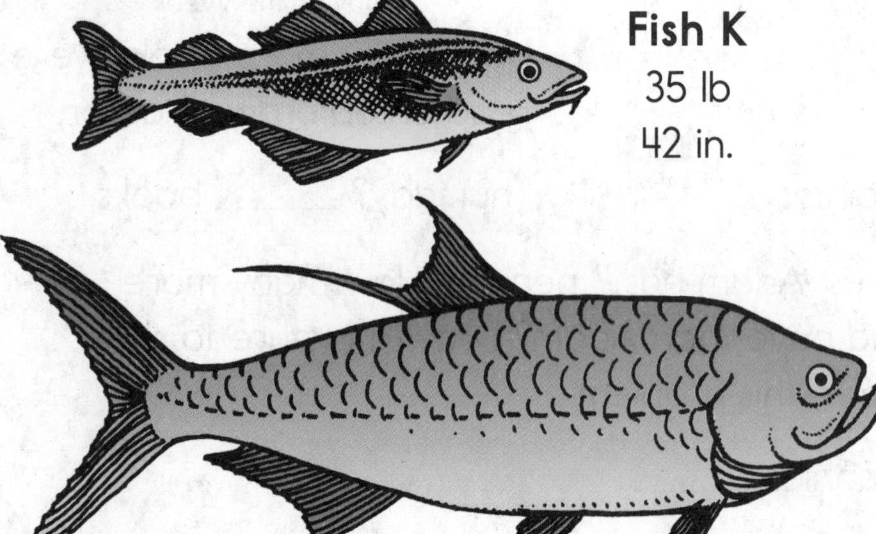

Fish L
100 lb
72 in.

"Fishy" Stories

Lesson 5-8

DATE

For each number story, do the following:

- Use the information on the Fish Poster to write the numbers you know in the change diagram.
- Write ? for the number you need to find or don't know.
- Write a number model with ? for the number you need to find.
- Solve the problem and write the answer.

1. Fish J swallows Fish B. How much does Fish J weigh now?

   Number model: ________________

   Answer: ______ pounds

   Start — Change → End

2. Fish K swallows Fish G. How much does Fish K weigh now?

   Number model: ________________

   Answer: ______ pounds

   Start — Change → End

Try This

3. Fish F swallows Fish B. How much does Fish F weigh now?

   Draw your own change diagram.

   Number model: ________________

   Answer: ______ pounds

Measuring Objects

Lesson 5-8

DATE

Look around the room. Find some objects you think are about **1 centimeter** long. Use a ruler to check the lengths. List some 1-centimeter objects here.	Now find some objects you think are about **1 inch** long. Use a ruler to check the lengths. List some 1-inch objects here.
Now find some objects you think are about **10 centimeters** long. Use a ruler to check the lengths. List some 10-centimeter objects here.	Now find some objects you think are about **10 inches** long. Use a ruler to check the lengths. List some 10-inch objects here.

Parts-and-Total Number Stories

Lesson 5-9

DATE

Lucy's Snack Bar Menu					
Sandwiches		**Drinks**		**Desserts**	
Hamburger	65¢	Juice	45¢	Apple	15¢
Hot dog	45¢	Milk	35¢	Orange	25¢
Ham and cheese	40¢	Lemonade	40¢	Banana	10¢
Peanut butter and jelly	35¢	Water	25¢	Cherry pie	40¢

For each problem, you are buying two items. Record the cost of each item in a diagram. Write ? for the number you need to find. Then write a number model and find the total cost of both items.

1. A lemonade and a banana

Total	
Part	Part

Number model:

____

Answer: ______¢

2. A hot dog and an apple

Total	
Part	Part

Number model:

____

Answer: ______¢

3. A hamburger and an orange

Draw your own parts-and-total diagram.

Number model:

____

Answer: ______¢

Math Boxes

Lesson 5-9

DATE

Math Boxes

1. You buy juice for 64¢. Use Ⓠ, Ⓓ, Ⓝ, and Ⓟ to show the coins you use to buy the juice.

2. Solve.

54 + 10 = ______

______ + 10 = 86

63 + ______ = 163

______ = 100 + 45

3. Put an X on the digit in the tens place.

456

309

510

4. Count up by 100s.

25, ______, 225, ______,

______, 525, 625

5. **Writing/Reasoning** What do you notice when you count up by 100 from 25 to 625 in Problem 4?

____

____

Temperature Changes

Lesson 5-10

DATE

Write ? in the End box. Then write a number model and find the answer. Fill in the thermometer to show the End number.

1

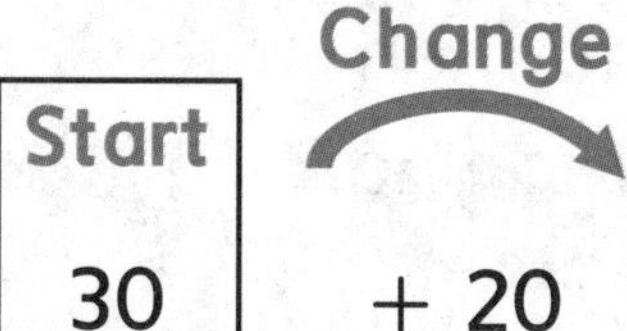

Number model:

____

Answer: ______°F

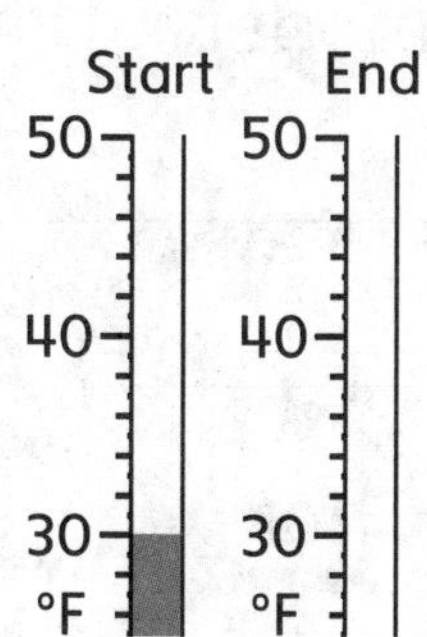

2

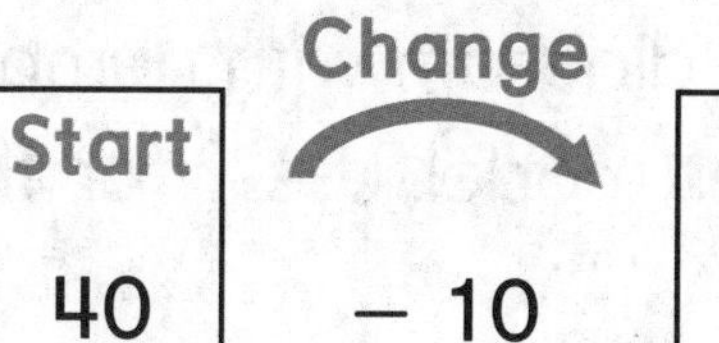

Number model:

____

Answer: ______°F

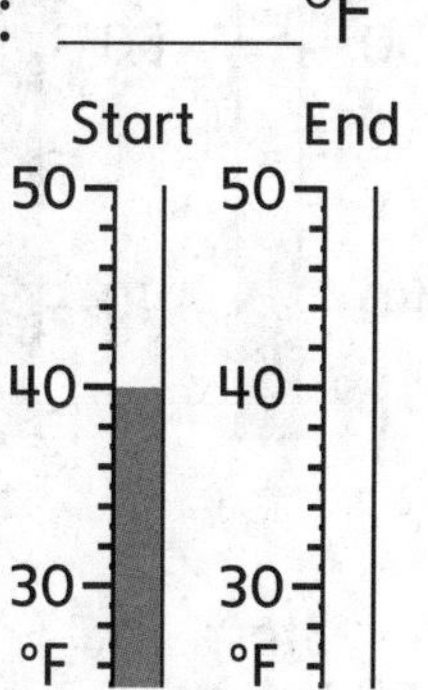

3

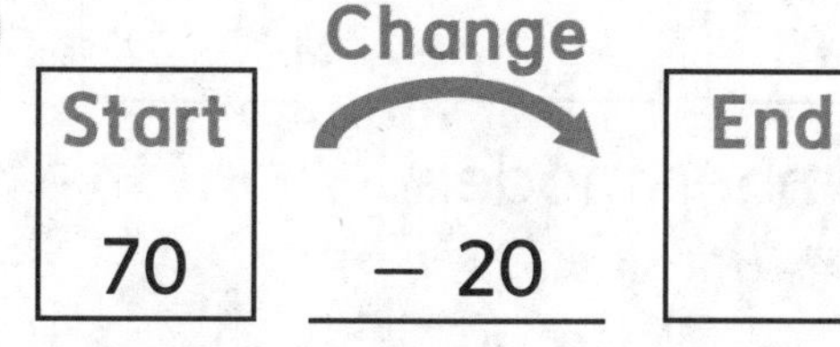

Number model:

____

Answer: ______°F

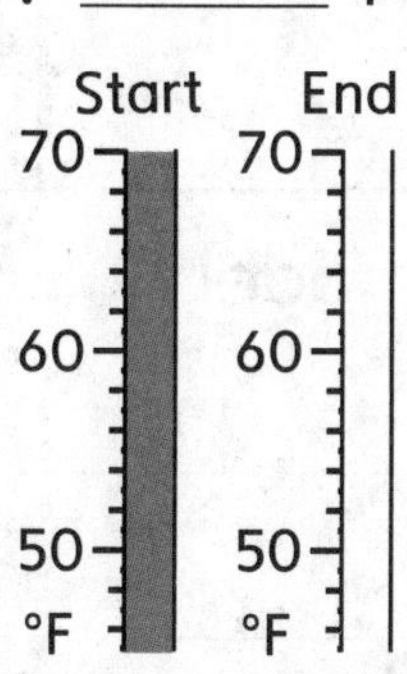

4

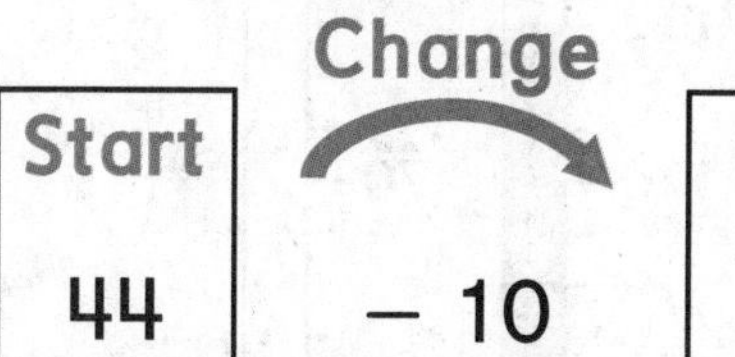

Number model:

____

Answer: ______°F

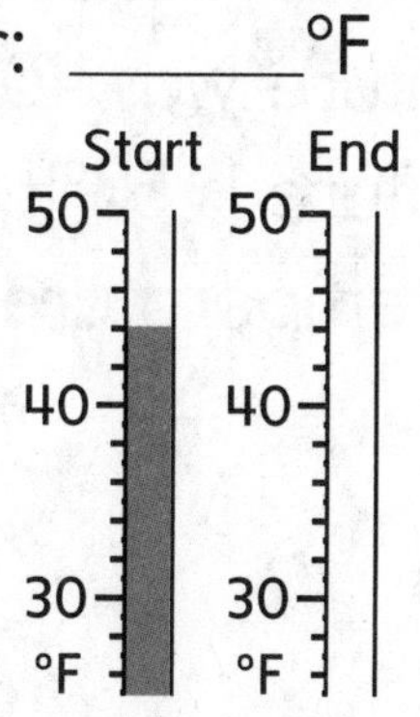

Temperature Changes (continued)

- Decide whether the change in temperature is a change to more or a change to less.
- Fill in the diagram with numbers from the problem. Then write a number model. Use ? for the number you want to find.
- Solve.

5

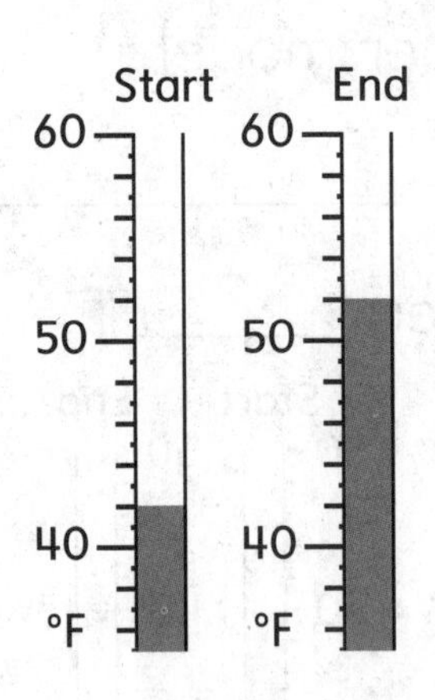

Change
Start
End

Number model: ______________________

Answer: ______ °F

6

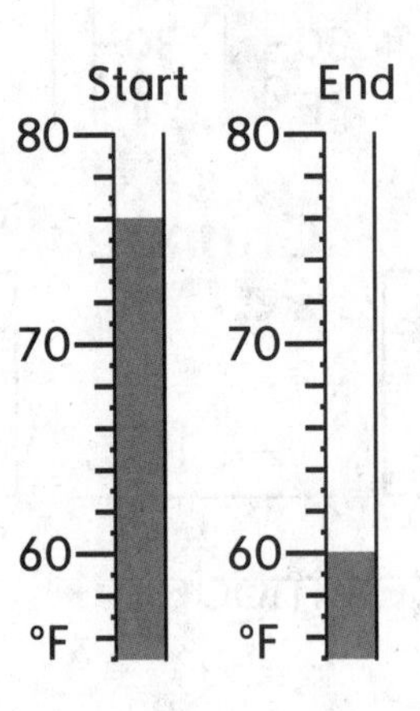

Change
Start
End

Number model: ______________________

Answer: ______ °F

7 When Isolde woke up, the temperature was 40°F. Now the temperature is 53°F. How much did the temperature change?

Change
Start
End

Number model: ______________________

Answer: ______ °F

Math Boxes

Lesson 5-10

DATE

Math Boxes

1 Use your centimeter ruler. Measure the length of this line segment:

Fill in the bubble next to the correct answer.

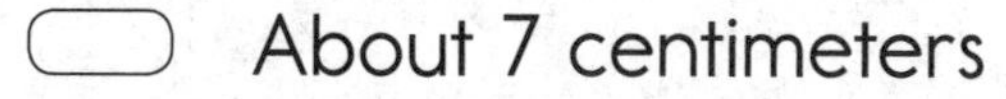

◯ About 7 centimeters

◯ About 6 centimeters

◯ About 8 centimeters

◯ About 10 centimeters

2 Write the time.

_____:_____

MRB 106–107

3 You buy an apple for 45¢. You pay with a dollar. How much change will you get?

_____¢

4 Write a number with 5 in the hundreds place, 0 in the ones place, and 3 in the tens place.

_____

5 You can make 45¢ with

1 Ⓠ **and** _____ Ⓓ.

You can make 60¢ with

3 Ⓓ **and** _____ Ⓝ.

6 Write the number word for 405.

____

Buying a Clock

Lesson 5-11

DATE

You buy a clock that costs $78. You pay with a $100 bill. How much is your change? You may use a number grid to help.

Answer: ______

Talk about your strategy with your partner.

Shopping Poster

Lesson 5-11

DATE

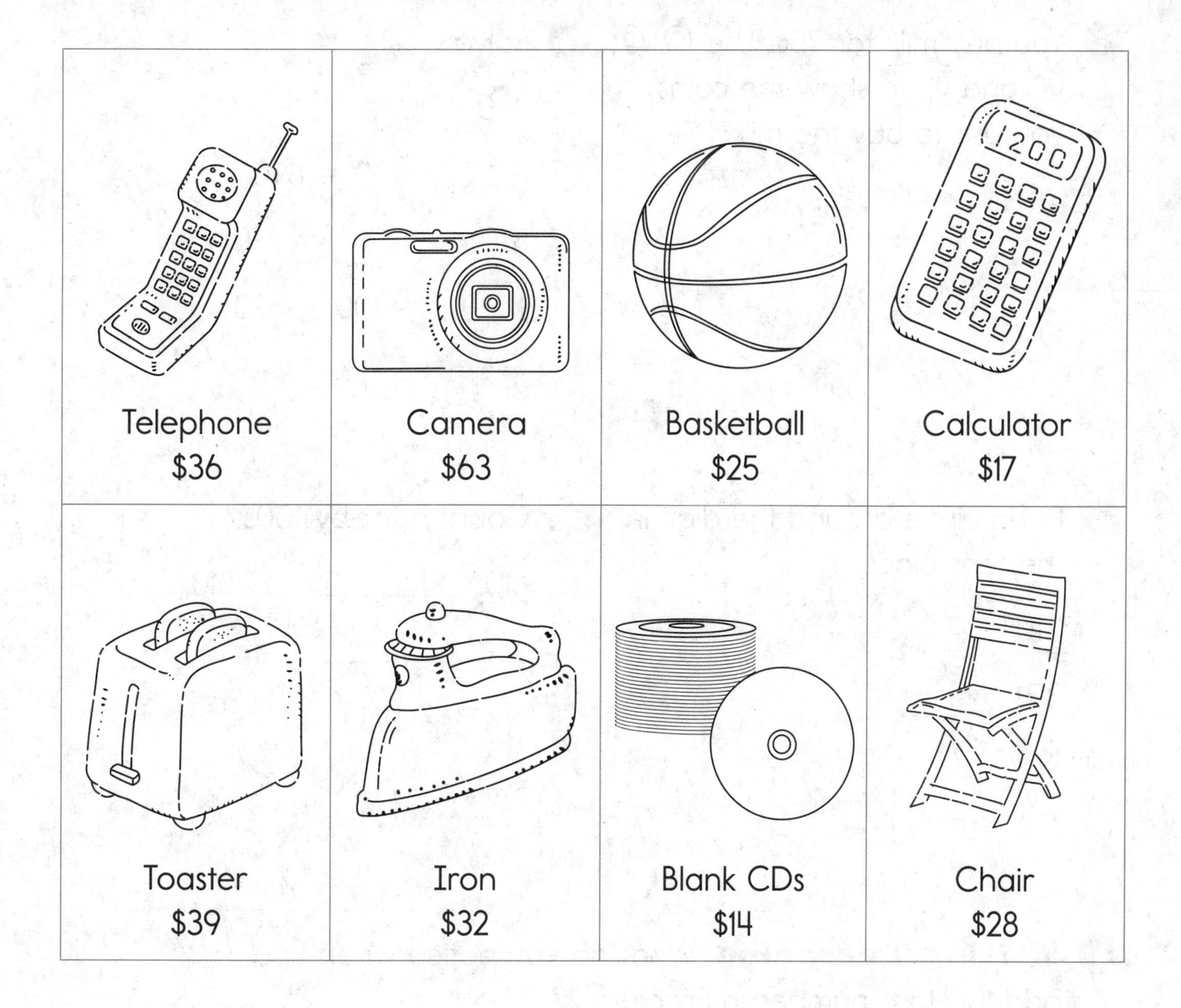

Telephone $36	Camera $63	Basketball $25	Calculator $17
Toaster $39	Iron $32	Blank CDs $14	Chair $28

Math Boxes

Lesson 5-11

DATE

Math Boxes

1 You buy milk for 75¢. Use Ⓠ, Ⓓ, Ⓝ, and Ⓟ to show the coins you use to buy the milk.

MRB 110-111

2 Solve.

330 + 10 = ______

______ + 10 = 890

570 + ______ = 670

______ = 100 + 620

3 Put a circle around the digit in the tens place.

210

391

199

MRB 73

4 Count back by 100s.

610, ______, ______, 310,

______, ______

5 **Writing/Reasoning** What do you notice when you add 100 to a number in Problem 2?

____

____

Math Boxes
Preview for Unit 6

Lesson 5-12

DATE

Math Boxes

1 Write each number in expanded form.

26 ________

54 ________

77 ________

92 ________

2 What number do the base-10 blocks show? ______

Use base-10 shorthand to show the number another way.

3 Last year a bamboo plant was 17 feet tall. It grew 10 feet this year. How tall is it now?

Number model:

____

Answer: ______ feet

MRB 27–28

4 What day did Molly swim the most laps?

____

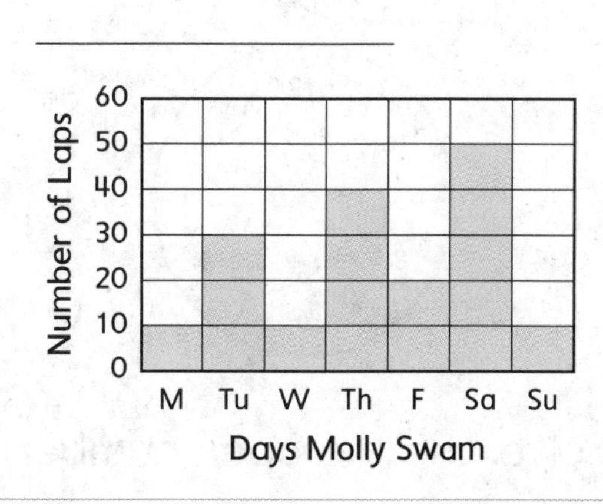

5 Sakura has 13 stickers. Aneko has 9 stickers. How many more stickers does Sakura have than Aneko? Draw a picture to show what you did to solve this problem.

Answer: ______ stickers

Math Boxes

Lesson 6-1

DATE

Math Boxes

1. Tony had 20 raisins. His sister gave him 15 more. How many raisins does he have now?

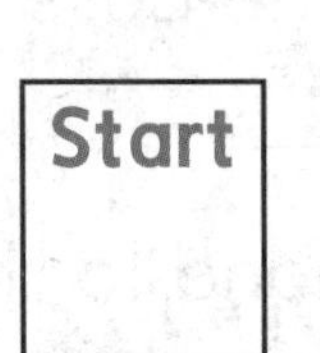

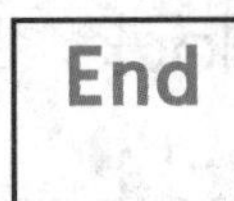
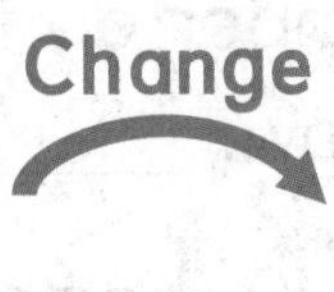

End

Number model:

____

Answer: ______ raisins

2. Measure the length of this leaf in centimeters. Circle the best answer.

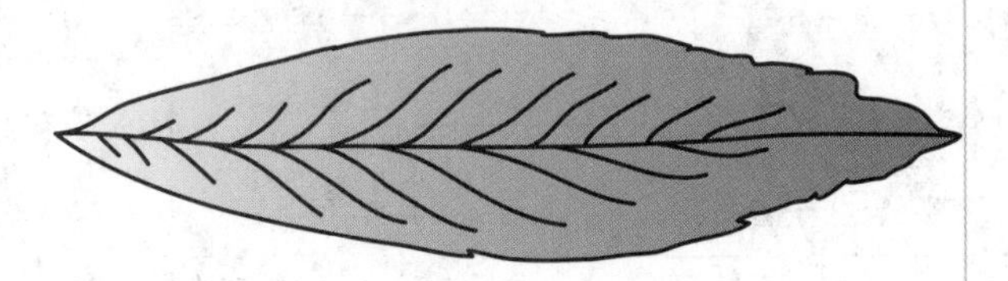

A. About 6 cm

B. About 7 cm

C. About 8 cm

D. About 10 cm

3. Draw the hands to show 8:15.

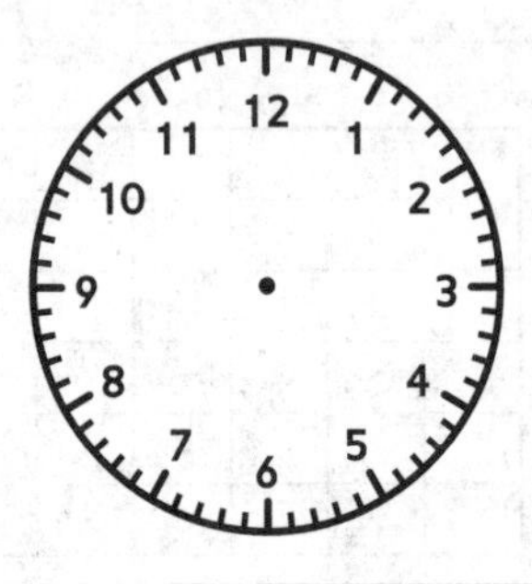

4. A bag of nuts costs 75¢. You pay with $1. How much change do you get?

______

5. Solve. Use the open number line to show what you did.

Rita has 25 buttons. Her sister gave her 20 more.
How many buttons does she have now? ______ buttons

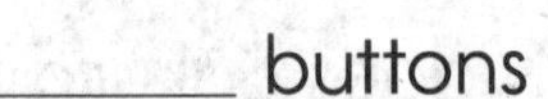

Making a Tally Chart

Lesson 6-1

DATE

Use tallies to show how many children in your class have each number of pockets.

Pockets	Children	
	Tallies	Number
0		
1		
2		
3		
4		
5		
6		
7		
8		
9		
10 or more		

Lesson 6-1

DATE

Drawing a Picture Graph

Draw a picture graph of the pockets data.

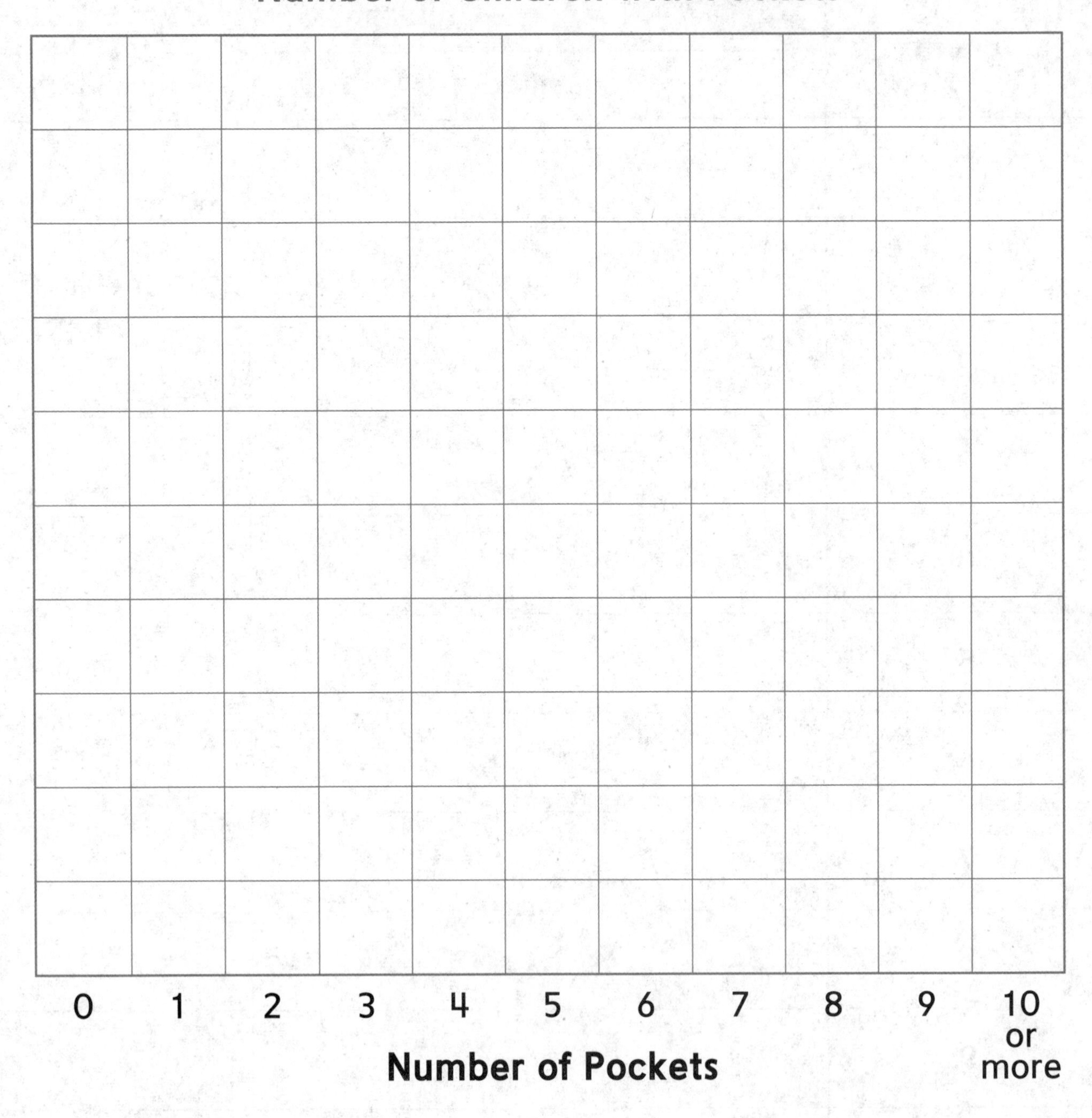

KEY: Each ☺ = 1 child

Drawing a Bar Graph

Lesson 6-1

DATE

Draw a bar graph of the pockets data.

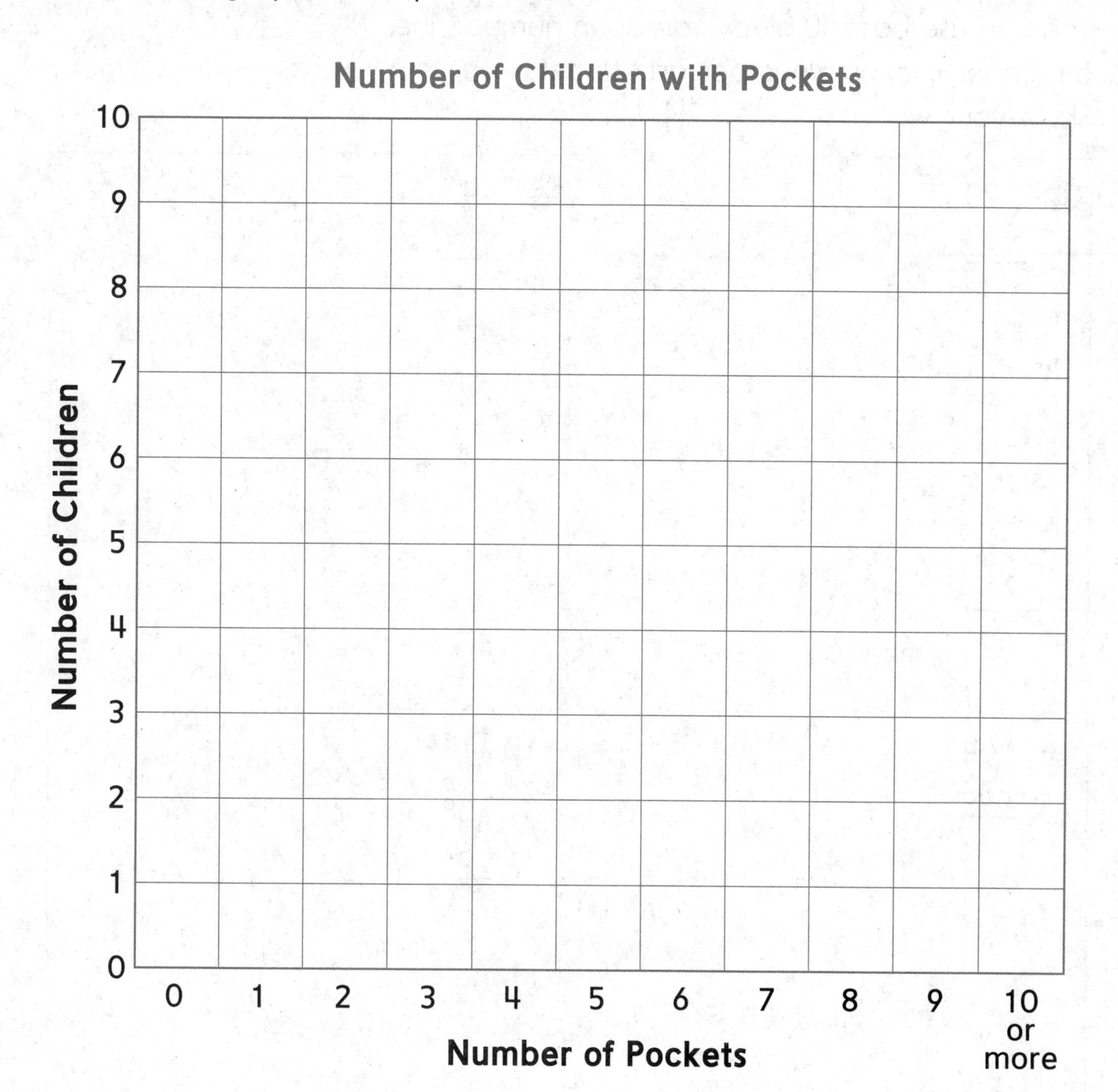

Lesson 6-1

DATE

Solving Addition Problems

Fill in the unit box. Then solve each problem.
You can use base-10 blocks, an open number line,
a number line, or a number grid to help you.
Show your work.

Unit

1. $\begin{array}{r} 20 \\ +63 \\ \hline \end{array}$

2. $\begin{array}{r} 42 \\ +39 \\ \hline \end{array}$

3. $\begin{array}{r} 54 \\ +38 \\ \hline \end{array}$

Try This

4. 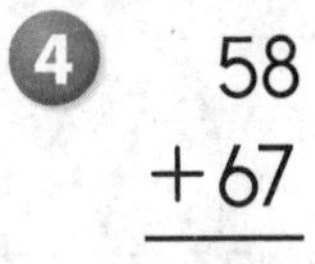

$\begin{array}{r} 58 \\ +67 \\ \hline \end{array}$

Comparison Number Stories

Lesson 6-2

DATE

For each number story, follow these steps:

- Write the numbers you know in the comparison diagram. Use ? for the number you need to find.
- Write a number model. Use ? for the number you don't know.
- Solve the problem and answer the question.

---

1 Barb's ribbon is 27 inches long. Cindy's ribbon is 10 inches long. Which ribbon is longer?

____

How much longer?

Number model:

____

Barb's ribbon is ______ inches longer than Cindy's.

Quantity

Quantity

______ Difference

---

2 Frisky lives on the 16th floor. Fido lives on the 7th floor. Who lives on the higher floor?

____

How many floors higher?

Number model:

____

Frisky lives ______ floors higher than Fido.

Quantity

Quantity

______ Difference

Comparison Number Stories

(continued)

Lesson 6-2

DATE

3. The green lizard is 36 inches long. The brown lizard is 21 inches long. Which lizard is longer?

____

How much longer?

Number model: _______________

The green lizard is _____ inches longer than the brown lizard.

Quantity

Quantity

Difference

---

4. A jacket costs $75. A pair of pants costs $20. Which costs less?

____

How much less?

Number model: _______________

The pants cost $_____ less than the jacket.

Quantity

Quantity

Difference

---

5. Jack scored 13 points. He scored 6 points more than Eli. How many points did Eli score?

Number model: _______________

Eli scored _____ points.

Quantity

Quantity

Difference

Math Boxes

DATE

1 Solve.

a. $53 + 10 =$ ______

b. $72 - 10 =$ ______

c. $192 + 10 =$ ______

d. ______ $= 301 - 10$

2 The temperature was 77°F in the afternoon. It was 60°F in the evening. How much did the temperature change?

Fill in the diagram and write a number model.

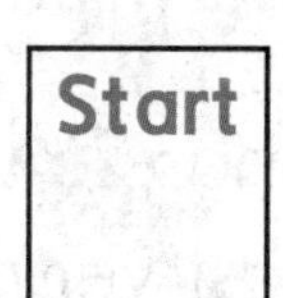

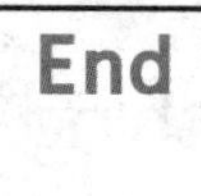

Number model: ____________

Answer: ______°F

3 Write the number word for 601.

____

4 Circle the expanded form that shows the larger number.

$700 + 10 + 8$

$600 + 90 + 9$

5 **Writing/Reasoning** Explain how you decided which number is larger in Problem 4.

____

____

____

Addition and Subtraction Number Stories

Lesson 6-3

DATE

Do the following for each number story:

- Write a number model. Use ? to show what you need to find. To help, you may draw a

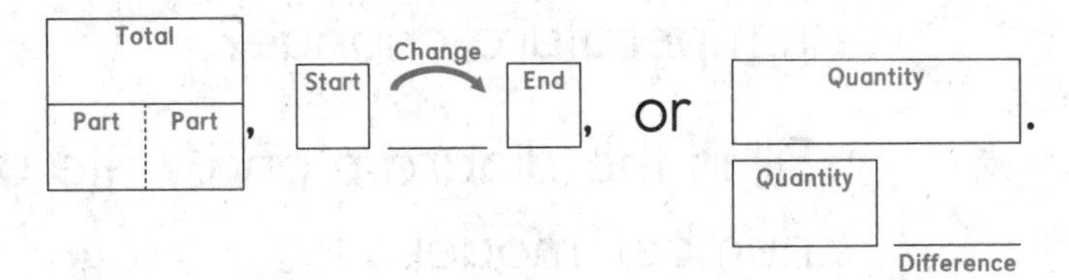

- Solve the problem and write the answer.

1. Rushing Waters now has 26 water slides. That is 9 more than last year. How many water slides were there last year?

   Number model: ______________________

   There were ______ water slides last year.

2. The Loop Slide is 65 feet high. The Tower Slide is 45 feet high. How much shorter is the Tower Slide?

   Number model: ______________________

   The Tower Slide is ______ feet shorter.

Addition and Subtraction Number Stories (continued)

Lesson 6-3

DATE

List two things you and your partner like to collect:

____________________ ____________________

Then do the following for each number story:

- Select an item from your collections list.
- Write a number model. Use ? to show the number you need to find. To help, you may draw a

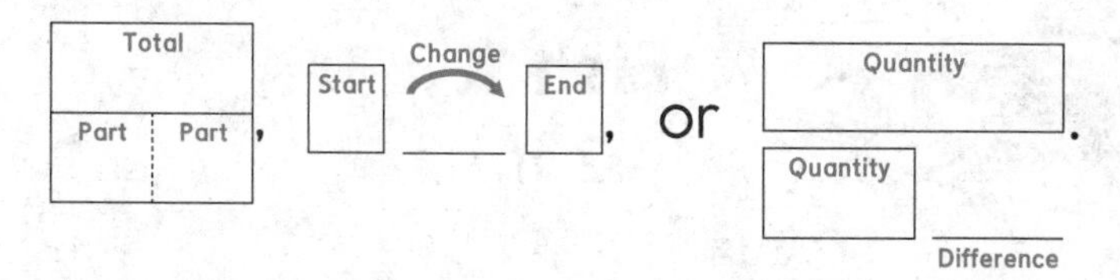

- Solve the problem and write the answer.

3. Colin has 20 ________________. Fiona has 30 ______________.

How many ________________ do they have in all?

Number model: ______________________

Colin and Fiona have ______ ________________ in all.
(answer) (unit)

4. Alexi had 34 ________________. He gave 12 ______________ to Theo.

How many ________________ does Alexi have now?

Number model: ______________________

Alexi has ______ ________________ now.
(answer) (unit)

Comparing Fish

Lesson 6-3

DATE

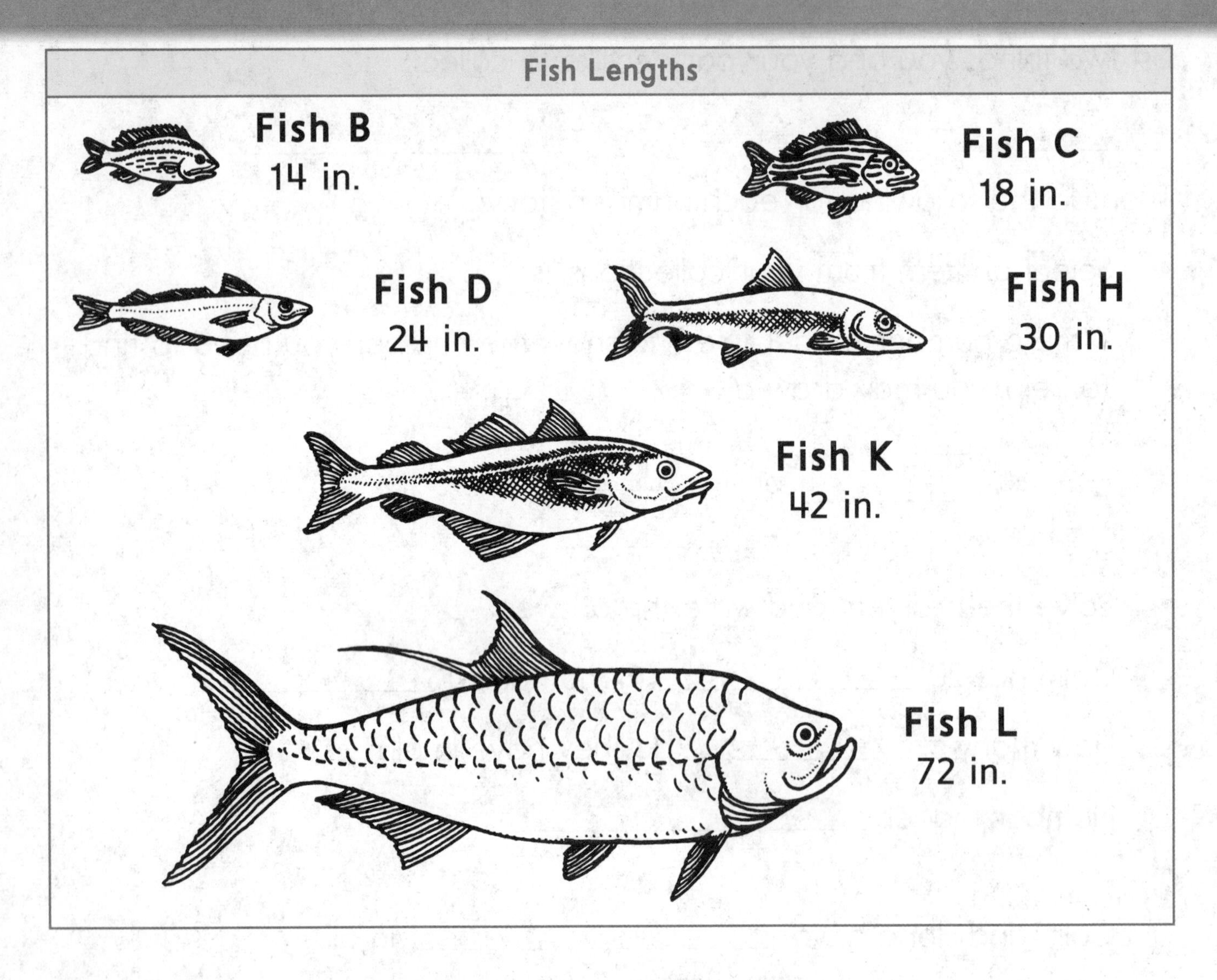

1. Fish C is ______ inches longer than Fish B.

Quantity
Fish C 18 in.

Quantity
Fish B 14 in.

______ Difference

2. Fish K is ______ inches longer than Fish H.

Quantity
Fish K ______ in.

Quantity
Fish H ______ in.

______ Difference

Comparing Fish (continued)

Lesson 6-3

DATE

3. Fish L is _____ inches longer than Fish H.

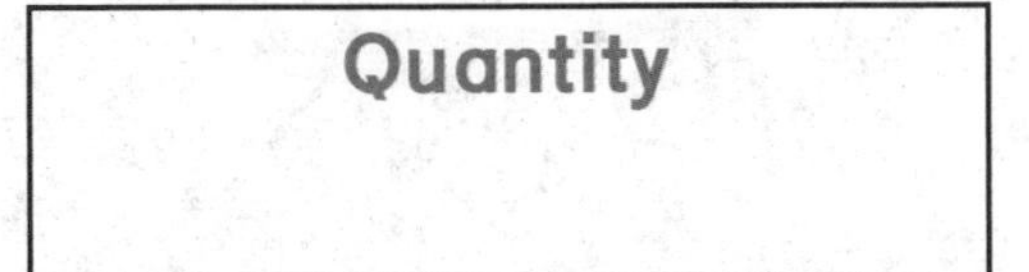

Quantity

_____ Difference

4. Fish D is _____ inches longer than Fish B.

Quantity

Quantity

_____ Difference

5. Fish H is 6 inches longer than _____.

Quantity

Quantity

_____ Difference

6. Fish L is 30 inches longer than _____.

Quantity

Quantity

_____ Difference

7. Fish C is 6 inches shorter than _____.

Quantity

Quantity

_____ Difference

8. Fish B is 16 inches shorter than _____.

Quantity

Quantity

_____ Difference

Math Boxes

1. One shelf has 24 books. The other shelf has 20 books. How many books are there in all?

Circle the correct answer.

A. 40 books

B. 42 books

C. 44 books

D. 50 books

2. Estimate the length of this bug in centimeters.

About ______ centimeters

Use your centimeter ruler to measure its length.

About ______ centimeters

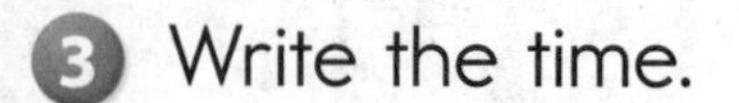

3. Write the time.

It is about ______:______.

4. You buy a pen for 72¢. You pay with $1. How much change do you get?

______

5. Solve. Use the open number line to show what you did.

There were 42 ducks at the pond. Then 21 more ducks joined them. How many ducks are there now? ______

DATE

Math Boxes

1 Solve.

a. 100 + 134 = ______

b. 201 − 100 = ______

c. ______ = 296 + 100

d. ______ = 407 − 100

2 The temperature was 50°F in the morning. It was 65°F in the afternoon. How much did the temperature change?

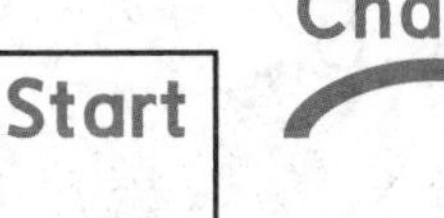
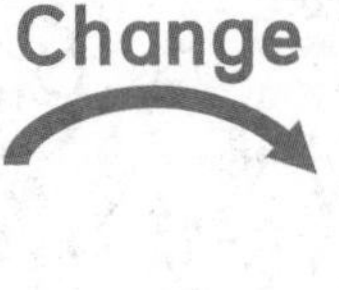

Number model:

____

Answer: ______°F

3 Write the number word for 911.

____

4 Circle the expanded form that shows the smaller number.

300 + 80 + 7

400 + 10 + 1

5 **Writing/Reasoning** Look at Problem 1. What pattern do you notice when you add 100? When you subtract 100?

____

____

____

Animal Heights and Lengths Poster

DATE

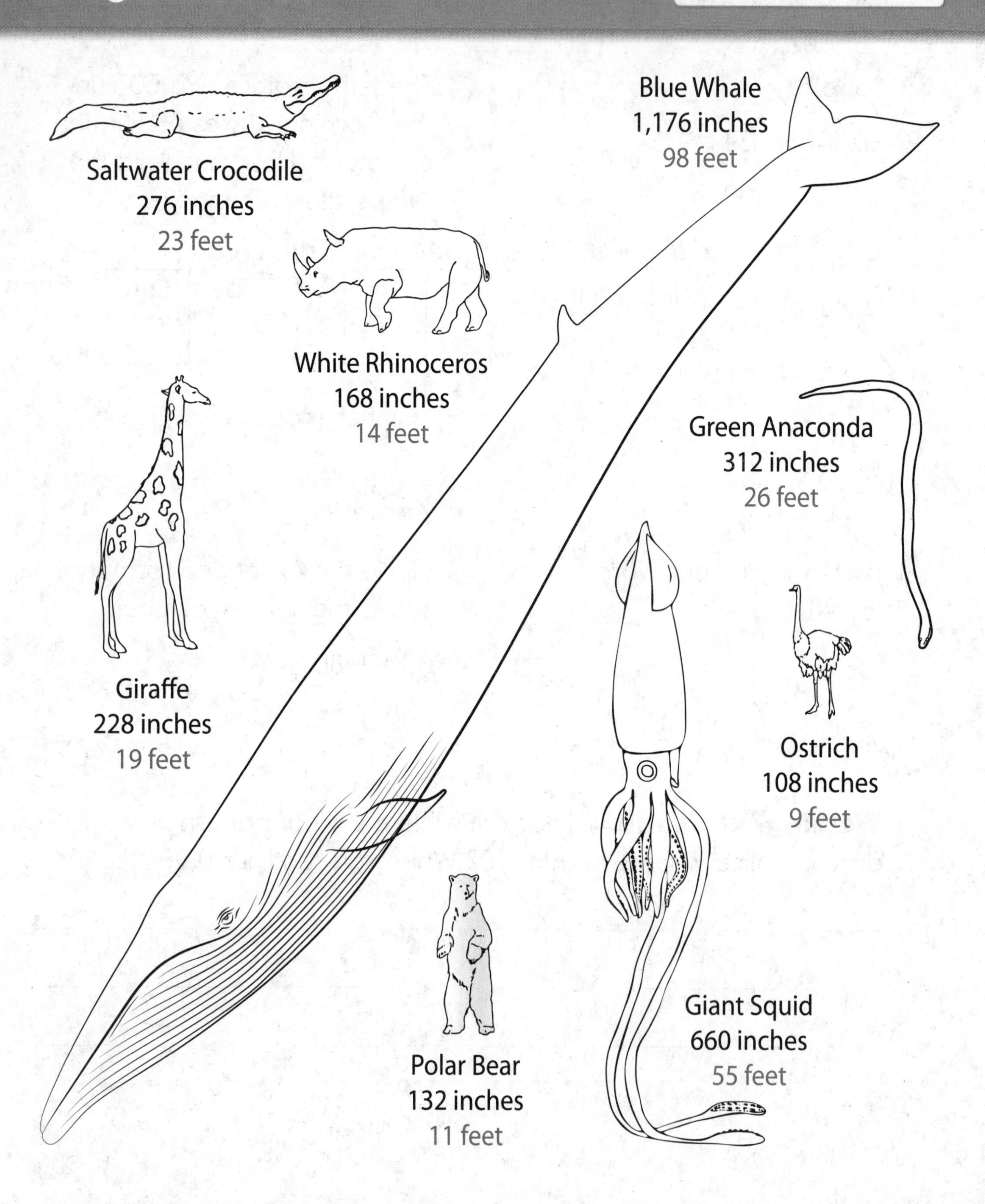

Silly Animal Stories

Lesson 6-4

DATE

Write and solve two number stories. Use information from the Animal Heights and Lengths Poster.

Example:

How much longer is the giant squid than the giraffe?

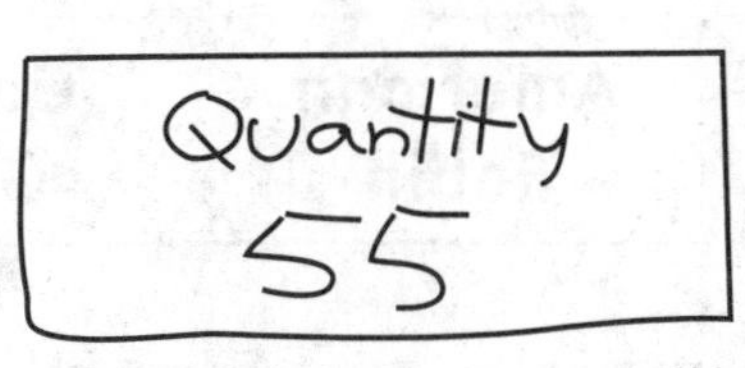

Unit
feet

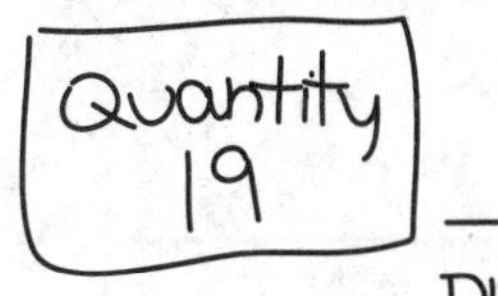

Number model: *55 − 19 = ?*

Answer: *The giant squid is 36 feet longer.*

1. ______

Unit

Number model: ______

Answer: ______

2. ______

Unit

Number model: ______

Answer: ______

Drawing a Bar Graph

Lesson 6-4

DATE

In one month a zoo recorded the number of eggs its birds laid. The number of eggs a bird lays at one time is called a "clutch." Use the data in the following table to draw a bar graph.

Type of Bird	American Robin	Canada Goose	Mallard Duck	Toucan
Number of Eggs in a Clutch	6	10	12	4

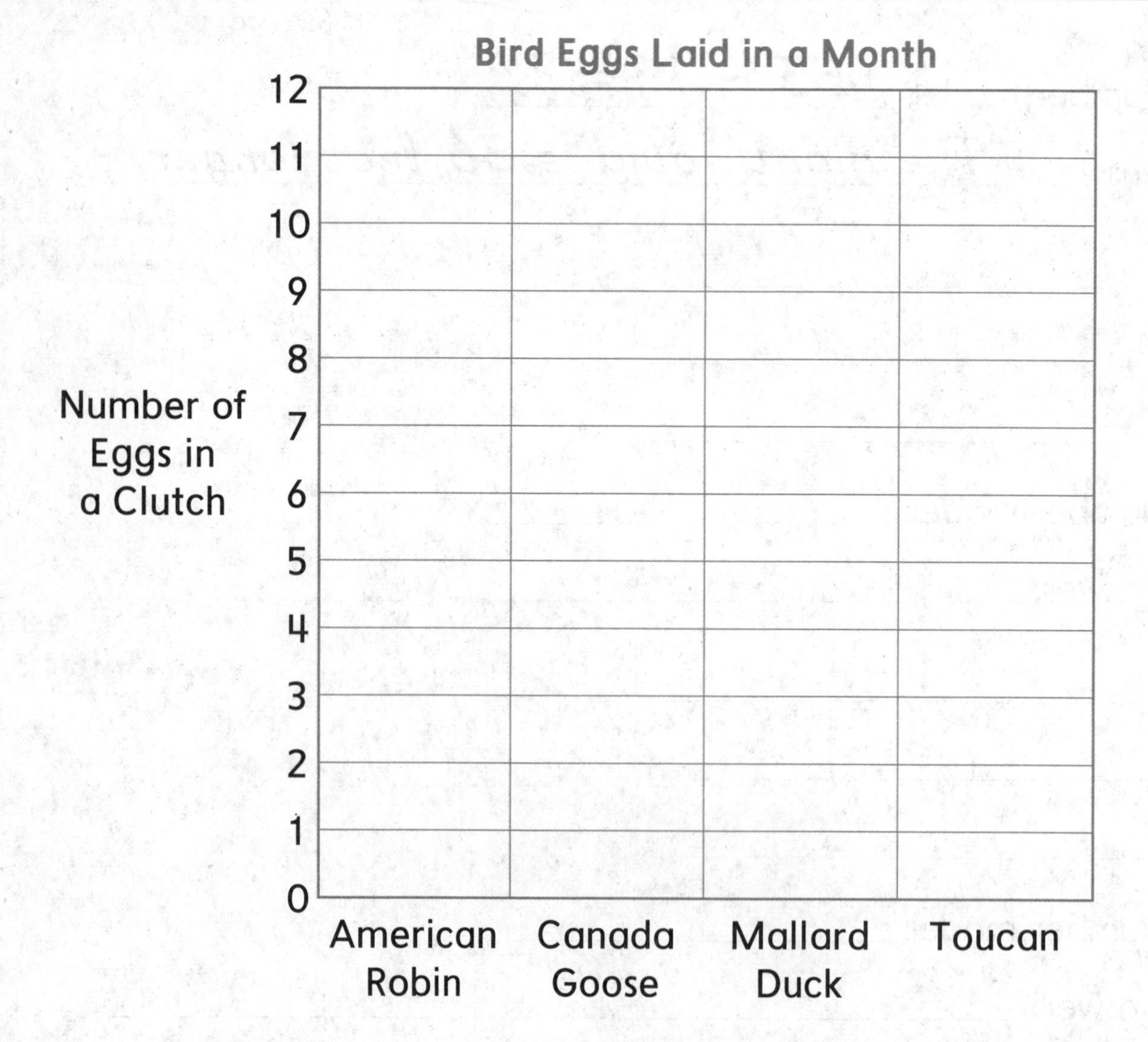

Drawing a Bar Graph (continued)

Lesson 6-4

DATE

Use the Bird Eggs Laid in a Month bar graph to solve the number stories.

1. The American robin and the toucan are tree birds. What is the total number of eggs laid by the tree birds?

   Number model: ____________________

   Answer: ______ eggs

2. The Canada goose and the mallard duck are water birds. What is the total number of eggs laid by the water birds?

   Number model: ____________________

   Answer: ______ eggs

3. How many more eggs did the mallard duck lay than the American robin?

   Number model: ____________________

   Answer: ______ eggs

4. Use the data in the bar graph to write your own number story.

   ____

   ____

   Number model: ____________________

   Answer: ______ eggs

Solving Two-Step Number Stories

Lesson 6-5

DATE

Do the following for each number story:

- Write a number model or number models.
  Use ? to show the number you need to find.
  To help, you may draw a

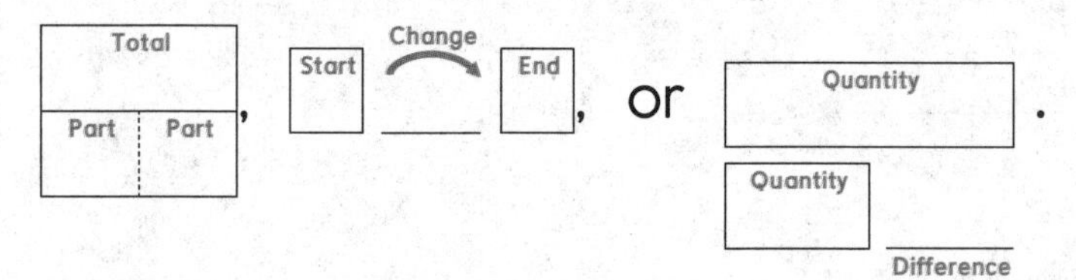

- Solve the problem and write the answer.

1. On Monday, Annabelle had 9 shells. On Tuesday, she found 12 more. On Wednesday, she gave 4 to her aunt. How many shells does Annabelle have now?

   Number model(s):

   ____

   Annabelle now has ______ shells.

2. Yvette bought 25 red balloons and 15 white balloons for a party. During the party 8 balloons popped. How many balloons did she have left when the party ended?

   Number model(s):

   ____

   Yvette had ______ balloons left.

Solving Two-Step Number Stories (continued)

Lesson 6-5

DATE

3. Tommy is playing a board game with his sister. On his first turn, he earned 15 points. On his second turn, he lost 7 points. On his third turn, he earned 12 points. How many points does he have now?

Number model(s): ______________________________

Tommy has _____ points.

4. Carrie had 23 markers. Luis gave her 7 more. She now has 15 more markers than Owen has. How many markers does Owen have?

Number model(s):

____

Owen has _____ markers.

Try This

5. On Monday Ellie had 18 gold stars. On Tuesday she got some more gold stars. On Wednesday she got 4 more gold stars. She now has 28 gold stars. How many gold stars did she get on Tuesday?

Number model(s):

____

Ellie got _____ gold stars on Tuesday.

DATE

Math Boxes

1. Draw a bar graph with this data: Jamal earned 3 stickers on Monday, 2 on Tuesday, and 4 on Wednesday.

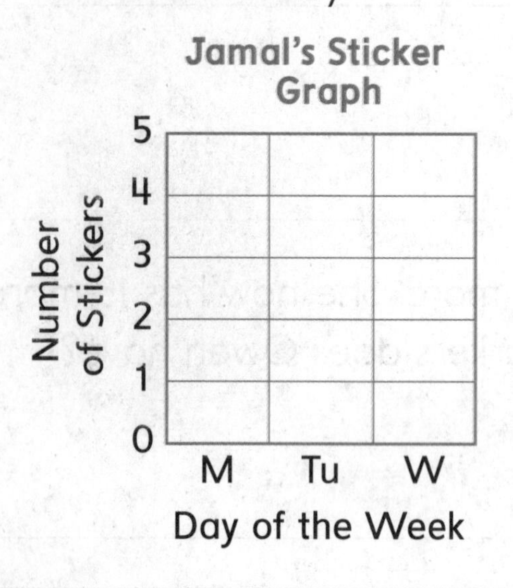

MRB 116

2. Luis rode his bike for 20 minutes on Monday and 25 minutes on Tuesday. How many minutes did he ride in all?

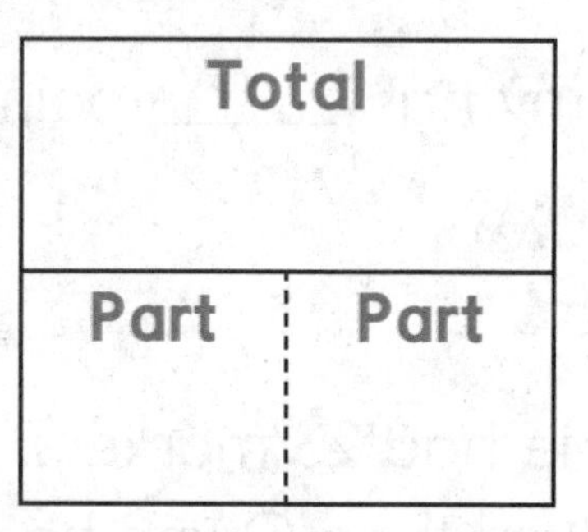

Number model:

____

Answer: ______ minutes

3. Write a number with 5 in the ones place, 3 in the hundreds place, and 7 in the tens place.

Answer: ______

4. Write the number that is 10 more.

196 ______ 301 ______

489 ______ 597 ______

5. Dale had 35 fish in his fish tank. He gave 20 fish away. How many fish did he have left?

Number model:

____

Answer: ______ fish

6. Ron's string is 45 inches long. Luke's string is 30 inches long. How much longer is Ron's string than Luke's?

Number model:

____

Answer: ______ inches

Estimating and Adding

Lesson 6-6

DATE

Fill in the unit box. For each problem:

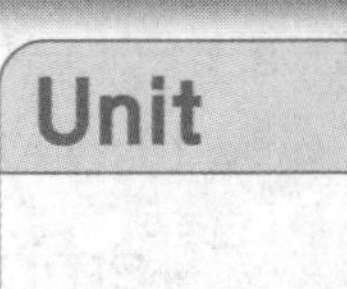

- Make a ballpark estimate.
- Solve the problem using any strategy you choose.
  Use words, numbers, or drawings to show your thinking.
- Explain how your estimate shows whether your answer makes sense.

1 32 + 26 = ? Ballpark estimate: ______________

Strategy:

32 + 26 = ______

Does your answer make sense? How do you know?

____

____

2 18 + 44 = ? Ballpark estimate: ______________

Strategy:

18 + 44 = ______

Does your answer makes sense? How do you know?

____

____

DATE

1. Use your Pattern-Block Template to draw a shape that has three sides.

MRB 122–125

2. 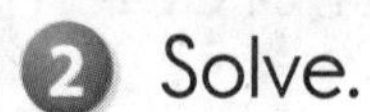 Solve.

   a. 217 + ______ = 227

   b. 10 + 483 = ______

   c. 507 = 497 + ______

   d. 409 = 10 + ______

3. Write each number in expanded form.

   69 ____________________

   24 ____________________

   345 ____________________

   180 ____________________

MRB 72–73

4. 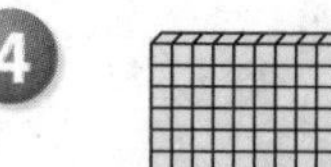
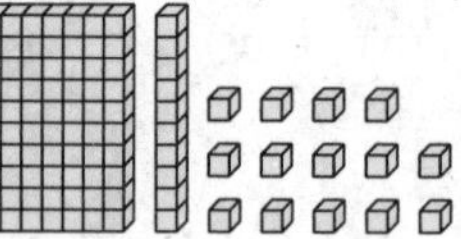

   What number do the base-10 blocks show? ______

   Use base-10 shorthand to show the number another way.

MRB 72–73

5. **Writing/Reasoning** Explain how you figured out the number shown by the base-10 blocks in Problem 4.

____

____

____

Addition with Base-10 Blocks

Lesson 6-7

DATE

Fill in the unit box.
Solve each problem using base-10 blocks.
Use base-10 shorthand to show what you did.
On the lines, record the partial sums and the answer.

Unit

Block	Flat	Long	Cube
Base-10 Shorthand	□	|	▪

Example:
$$\begin{array}{r} 23 \\ +\ 46 \\ \hline \end{array}$$

|| ▪▪▪
|||| ▪▪▪▪▪▪

Answer: 60 + 9 = 69

1.
$$\begin{array}{r} 41 \\ +\ 35 \\ \hline \end{array}$$

Answer: _____ + _____ = _____

2.
$$\begin{array}{r} 67 \\ +\ 38 \\ \hline \end{array}$$

Answer:

_____ + _____ = _____

3.
$$\begin{array}{r} 123 \\ +\ 128 \\ \hline \end{array}$$

Answer:

_____ + _____ + _____ = _____

Number Stories

DATE

For each number story:

- Write a number model. Use ? to show the number you need to find. To help, you may draw a

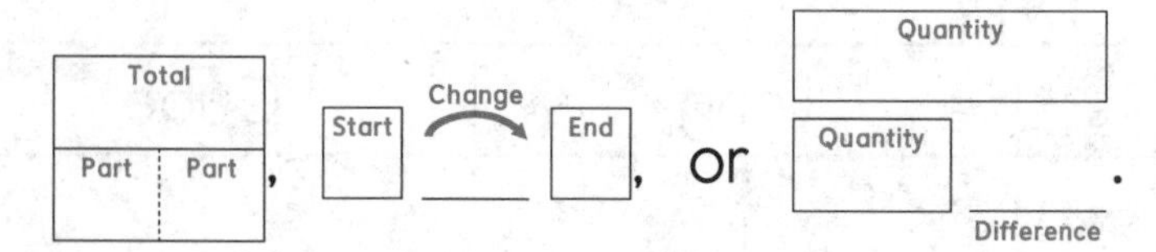

- Solve the problem and write the answer. Include the unit.

1. Jack drove 40 miles to a gas station. Then he drove 30 miles from the gas station to his friend's house. How many miles did Jack drive in all?

   Number model: ____________________

   Answer: ____________________

2. Emma found two leaves. One leaf was 9 centimeters longer than the other. The longer leaf was 20 centimeters long. How long was the shorter leaf?

   Number model: ____________________

   Answer: ____________________

Try This

3. A fish weighs 35 pounds. An octopus weighs 20 pounds. A crab weighs 2 pounds. How much do all three weigh together?

   Number model: ____________________

   Answer: ____________________

Math Boxes
Preview for Unit 7

1 Solve.

Unit
pencils

14 + ______ = 20

20 = 12 + ______

11 + ______ = 20

______ + 13 = 20

2 Meg has 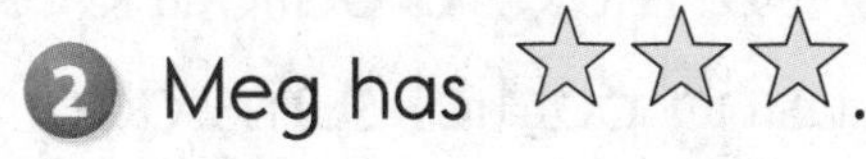.

Dan has ☆☆☆☆☆☆☆.

Jen has ☆☆.

How many stars are there in all? ______ stars

3 Name something that is about 1 foot long.

____

4 Erin can jump about 1 foot high. John can jump about 2 feet high. How much higher can John jump?

About ______ ft

5

Books Read

Number of Books

8
6
4
2
0

Juan
Lilly
Grace
Terell

Who read the most books? ______________

Who read the fewest books? ______________

6 Write a number story to match the number model.

4 + 6 − 5 = ?

____

____

____

____

____

Partial-Sums Addition

Lesson 6-8

DATE

Unit

For Problems 1–3, make a ballpark estimate. Then solve the problem using partial-sums addition. Show your work. Use your estimate to check that your answer makes sense.

Example: 59 + 26 = ?

Ballpark estimate:

60 + 30 = 90

Think:
50 + 20 =
9 + 6 =

$$\begin{array}{r} 59 \\ +\ 26 \\ \hline 70 \\ 15 \\ \hline 85 \end{array}$$

Think:
50 + 9
20 + 6

1. Ballpark estimate:

____

$$\begin{array}{r} 34 \\ +\ 71 \\ \hline \end{array}$$

2. Ballpark estimate:

____

$$\begin{array}{r} 136 \\ +\ 157 \\ \hline \end{array}$$

3. Ballpark estimate:

____

$$\begin{array}{r} 122 \\ +\ 53 \\ \hline \end{array}$$

4. Solve one of the problems a different way. Explain your strategy.

____

____

DATE

Math Boxes

1 How many more books did Alex read in Week 4 than in Week 1? _____

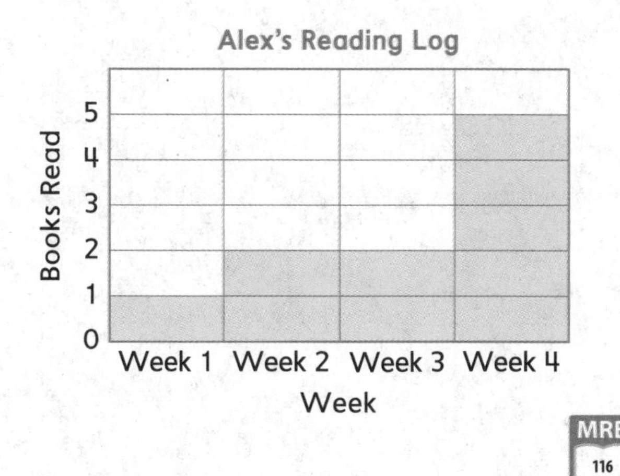

MRB 116

2 Ali had 35¢ and then found 40¢. How much money does he have in all? Fill in the diagram and write a number model.

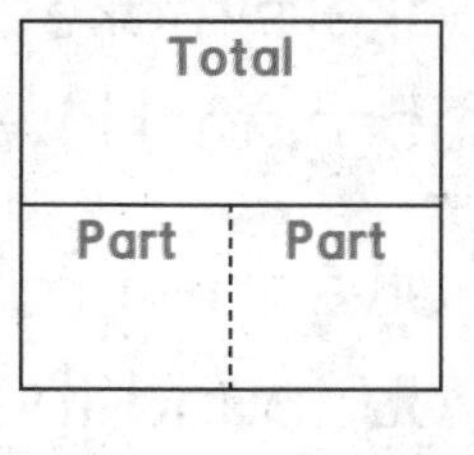

Number model:

____

Ali has _____.

MRB 25–26

3 What number do the base-10 blocks show? _____

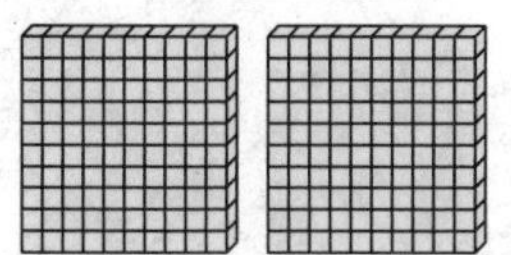

MRB 73

4 Which number is 10 less than 301? Circle the correct answer.

A. 281 **B.** 311

C. 291 **D.** 310

5 The store sold 20 T-shirts on Tuesday. It sold more on Wednesday. It sold 42 T-shirts in all. How many did it sell on Wednesday?

Number model:

____

Answer: _____ T-shirts

MRB 27–28

6 Write a number model and solve.

Ann's arm span is 48 inches. Bell's arm span is 60 inches. How much longer is Bell's arm span than Ann's?

Number model:

____

Answer: _____ inches

MRB 30–31

A Subtraction Number Story

DATE

Bakers make 56 loaves of bread. They sell 24 loaves in the morning. How many loaves are left for sale in the afternoon?

Solve the problem using any strategy. Be ready to explain how you found the answer.

Number model: ______________________ loaves

Math Boxes

DATE

1. Use your Pattern-Block Template to draw a shape that has four sides.

2. 317 + ______ = 327

   10 + 673 = ______

   708 = 698 + ______

   804 = 10 + ______

3. Write each number in expanded form.

   29 ______________________

   53 ______________________

   134 ______________________

   300 ______________________

4. 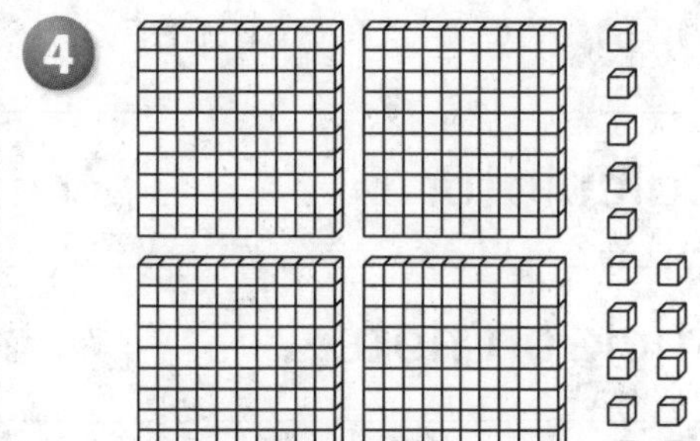

   What number do the base-10 blocks show? ______

   Use base-10 shorthand to show the number another way.

5. **Writing/Reasoning** How did you find the second way to show the number in Problem 4? Explain.

____

____

Comparing Lengths

Lesson 6-10

DATE

Get an eraser, a calculator, a stick-on note, and *My Reference Book*. Measure each object's length to the nearest inch. Record the measures.

Eraser: about ______ inches

Calculator: about ______ inches

My Reference Book: about ______ inches

Stick-on note: about ______ inches

Circle the name of the longer object.	Find the difference in length between the two objects.
Eraser Calculator	About ______ inches
Eraser Stick-on note	About ______ inches
Eraser *My Reference Book*	About ______ inches
My Reference Book Calculator	About ______ inches
My Reference Book Stick-on note	About ______ inches
Stick-on note Calculator	About ______ inches

Which object is the longest? ____________________

Which object is the shortest? ____________________

Explain how you found the differences in length between pairs of objects.

____

____

Math Boxes

1. On a nature walk Amy saw 5 snails, 2 turtles, and 6 caterpillars. Draw the graph.

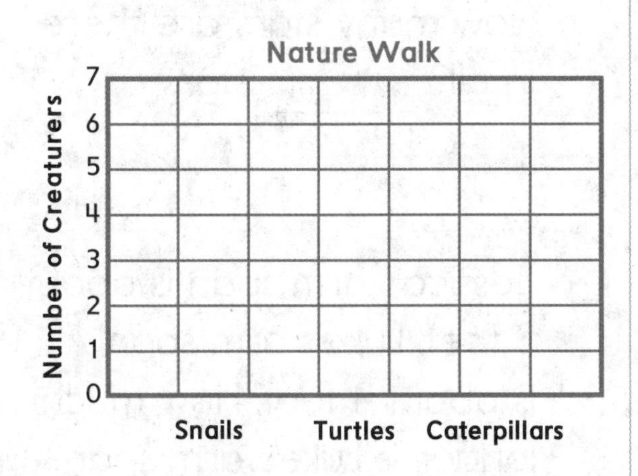

2. Julie had 30 crayons. She bought 18 more. How many does she have in all?

Start — Change — End

Number model:

____

Answer: ______ crayons

3. Which number has a 4 in the ones place, an 8 in the tens place, and a 7 in the hundreds place? Circle the answer.

A. 784 **B.** 874

C. 487 **D.** 748

4. Write the number that is 100 more.

456 ______ 402 ______

718 ______ 390 ______

5. Devon has some stickers. His friend gives him 20 more. Now he has 33 stickers. How many did he have to start?

Number model:

____

Answer: ______ stickers

6. The oak tree is 38 feet tall. The maple tree is 20 feet tall. How much shorter is the maple tree?

Number model:

____

Answer: ______ feet

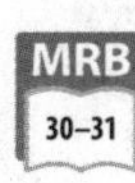

Math Boxes Preview for Unit 7

1 Solve.

Unit
counters

24 + ______ = 30

30 = 22 + ______

21 + ______ = 30

______ + 23 = 30

2 Tim had ☆☆☆☆.
Roy had ☆☆☆.
Don had ☆☆☆☆☆☆.
How many stars are there in all? ______ stars

3 Name something that is about 3 feet long.

____

4 Jessica's arm span is about 3 feet. Luke's arm span is about 4 feet. How much longer is Luke's arm span?

About ______ ft

5 How many brothers and sisters does Ben have?

______

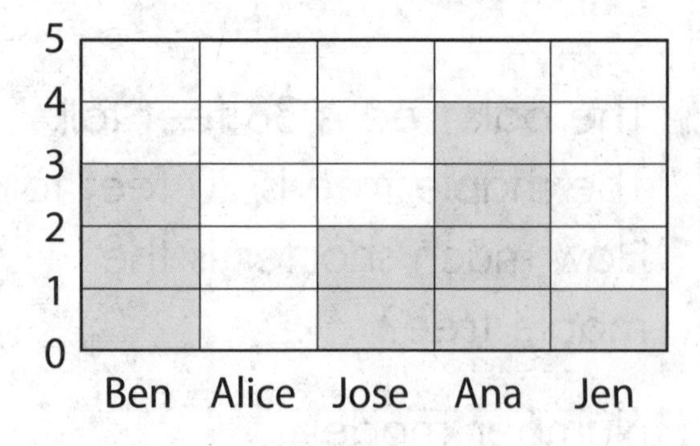

6 Write a number story to match the number model.
13 + 7 + 6 = ?

____

____

____

____

____

____

____

Hit the Target Record Sheet

Lesson 7-1

DATE

Record three rounds of *Hit the Target.*

Example Round

Target number: 40

Starting Number	Change	Result	Change	Result	Change	Result
12	+ 38	50	− 10	40		

Round 1

Target number: ______

Starting Number	Change	Result	Change	Result	Change	Result

Round 2

Target number: ______

Starting Number	Change	Result	Change	Result	Change	Result

Round 3

Target number: ______

Starting Number	Change	Result	Change	Result	Change	Result

Bamboo Plant Number Stories

Lesson 7-1

DATE

Bamboo is one of the world's fastest-growing plants. Some types of bamboo grow more than 24 inches per day and reach heights close to 100 feet! For one week a growing bamboo plant was measured. The chart below shows its height at the beginning of each day.

Bamboo Plant Growth for One Week

Sun.	Mon.	Tues.	Wed.	Thurs.	Fri.	Sat.
12 in.	26 in.	40 in.	57 in.	63 in.	80 in.	99 in.

Use the information above to solve the following number stories.

1. How many inches did the bamboo plant grow from Tuesday to Friday?

   Number model:

   ____

   Answer: ______ inches

2. How many inches did the bamboo plant grow from Thursday to Friday?

   Number model:

   ____

   Answer: ______ inches

Bamboo Plant Number Stories
(continued)

Lesson 7-1

DATE

3. How many inches taller was the bamboo plant on Saturday than on Tuesday?

Number model:

____

Answer: _____ inches

4. How many inches taller was the bamboo plant on Saturday than on Sunday?

Number model:

____

Answer: _____ inches

5. Make up and solve your own number story about the bamboo plant.

____

____

____

Number model: _______________

Answer: _____ inches

Math Boxes

1 Solve.

Unit

214 + 100 = ______

______ + 115 = 215

591 + ______ = 691

______ = 100 + 218

2 Solve.

$$\begin{array}{r} 23 \\ +\ 58 \\ \hline \end{array}$$

Ballpark estimate:

____

3 Miguel had 18 grapes. He ate 7. His mother gave him 10 more. How many grapes does he have now?

Number model(s):

____

Answer: ______ grapes

4 Fill in the missing numbers.

	908
917	

5 **Writing/Reasoning** How might the ballpark estimate in Problem 2 help you with the problem?

____

____

____

Counting Pencils

DATE

Lia has 3 pencils. Thomas has 6 pencils. Nate has 7 pencils. How many pencils do they have in all?

Solve the problem using any strategy. Share your strategy with your partner.

Number model: ______________________ pencils

Math Boxes

DATE

1 Write a number model. You may draw a diagram to help. Solve the problem.

Hank is 46 years old. Tom is 16 years younger than Hank. How old is Tom?

Number model:

____

Answer: ______

2 Which day had 25 minutes of recess?

____

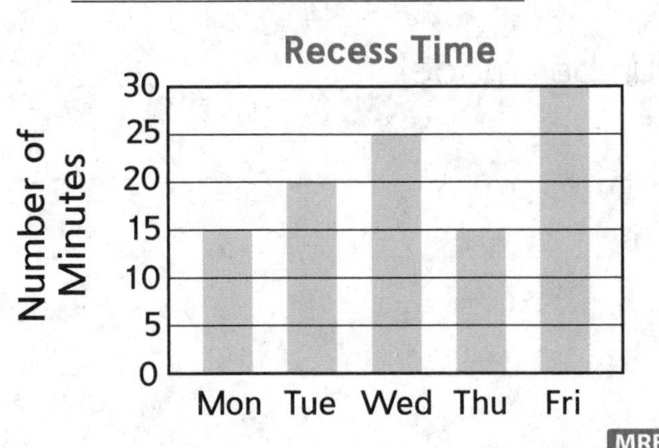

3 Measure the length of the line in both centimeters and inches.

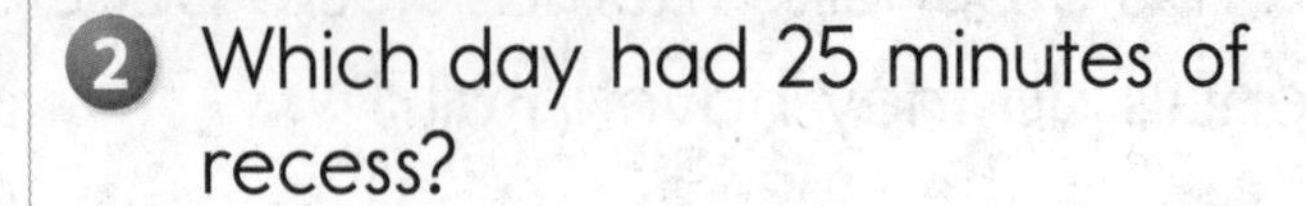

about ______ cm

about ______ in.

4 Write a number model. You may use a diagram to help. Solve the problem.

Arlie harvested 12 bushels of corn and 19 bushels of tomatoes. How many bushels in all?

Number model:

____

Answer: ______ bushels

5 Write a number story to match $22 + 20 = ?$. Then solve.

____

____

Basketball Addition

Lesson 7-3

DATE

	Points Scored			
	Team 1		Team 2	
	1st Half	2nd Half	1st Half	2nd Half
Player 1				
Player 2				
Player 3				
Player 4				
Player 5				
Team Score				

Point Totals	**1st Half**	**2nd Half**	**Final**
Team 1	_____	_____	_____
Team 2	_____	_____	_____

1. Which team won the first half? ____________________

   By how much? _____ points

2. Which team won the second half? ____________________

   By how much? _____ points

3. Which team won the game? ____________________

   By how much? _____ points

Drawing a Picture Graph

Lesson 7-3

DATE

An *astronaut* is a person who is trained to fly into space. This table shows how many trips four American astronauts took into space. Use the information to draw a picture graph.

Astronaut Name	Curtis Brown	Bonnie Dunbar	Jerry Ross	James Wetherbee
Number of Trips into Space	6	5	7	6

Trips Astronauts Took into Space

Curtis Brown	Bonnie Dunbar	Jerry Ross	James Wetherbee

KEY: ◯ = 1 trip into space

Drawing a Picture Graph
(continued)

Use the information from the picture graph on page 172 to solve the number stories. You may draw a diagram to help.

1. How many trips in all did Bonnie Dunbar and James Wetherbee take into space?

   Number model: ____________________

   They took _____ trips into space.

2. How many trips in all did the male astronauts take into space? (Bonnie Dunbar is the only female astronaut in the table.)

   Number model: ____________________

   The male astronauts took _____ trips into space.

3. How many more times did Jerry Ross go into space than Bonnie Dunbar?

   Number model: ____________________

   Jerry Ross went into space _____ more times.

4. Use the information from the picture graph to write your own number story.

   ____

   ____

   ____

Math Boxes

Lesson 7-3

DATE

1 Solve.

Unit

445 + 100 = ______

______ + 342 = 442

778 + ______ = 878

______ = 100 + 265

2 Solve.

$$\begin{array}{r} 36 \\ +\ 49 \\ \hline \end{array}$$

Ballpark estimate:

3 There are 12 children in the pool. Then 5 get out. Then 9 more jump in. How many children are in the pool now?

Number model(s):

______

Answer: ______ children

4 Fill in the missing numbers.

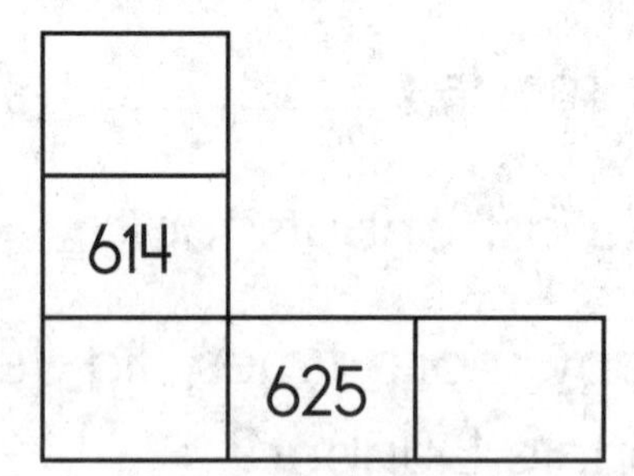

5 **Writing/Reasoning** Karen wrote the number model 12 + 5 + 9 = ? for Problem 3. Do you agree? Explain your answer.

______

______

______

MRB 34–35

Personal References Hunt

Lesson 7-4

DATE

1 Find things that are about 1 inch long, about 1 foot long, and about 1 yard long. Use a ruler, a tape measure, or a yardstick to help you. List them below. You can use these as your personal references to help you estimate lengths.

My Personal References		
About 1 in.	About 1 ft	About 1 yd

2 Find things that are about 1 centimeter long, about 10 centimeters long, and about 1 meter long. Use a ruler, a tape measure, or a meterstick to help you. List them below. You can use these as your personal references to help you estimate lengths.

My Personal References		
About 1 cm	About 10 cm	About 1 m

Yards

Lesson 7-4

DATE

For Problems 1–3, do the following:

- Choose a distance.
- Estimate the distance in yards.
- Measure the distance to the nearest yard.
- Compare your measurement with your estimate. Talk to your partner about why they might be different.

Distance	My Estimate	My Yardstick Measurement
Example: *From the door to the window*	About 5 yards or Between ____ and ____ yards	About ____ yards or Between 8 and 9 yards
1 ________	About ____ yards or Between ____ and ____ yards	About ____ yards or Between ____ and ____ yards
2 ________	About ____ yards or Between ____ and ____ yards	About ____ yards or Between ____ and ____ yards
3 ________	About ____ yards or Between ____ and ____ yards	About ____ yards or Between ____ and ____ yards

Math Boxes

DATE

1 Solve. You can draw a diagram to help.

A coat costs $65. Shoes cost $30. How much less do the shoes cost?

Number model: ______________

The shoes cost $______ less.

MRB 30–31

2 How many baskets did Anna and Kim make all together?

How Many Baskets?

Number of Baskets: 0, 1, 2, 3, 4, 5, 6, 7

Anna Jane Kim Carla

Ⓐ 5 baskets

Ⓑ 3 baskets

Ⓒ 2 baskets

Ⓓ 8 baskets

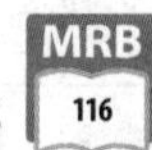

3 Measure the length of this line segment:

About ______ cm

About ______ in.

4 Oliver picked 16 tulips and 20 daffodils. How many flowers did he pick in all?

Number model: ______________

Answer: ______ flowers

5 Write a number story to match this number model: $45 - 20 = ?$

____

____

Solve your number story. Answer: ______________

Estimating and Measuring

Lesson 7-5

DATE

For each length:

- Choose one metric unit: centimeters (cm) or meters (m).
- Use your personal references to estimate the length. Record your estimate. Be sure to write the unit.
- Choose a measuring tool and measure the length. Record your measurement. Be sure to write the unit.
- Choose one U.S. customary unit: inches (in.), feet (ft), or yards (yd). Repeat the steps again using that unit.

Lengths	Metric Units		U.S. Customary Units	
	Estimate	Measurement	Estimate	Measurement
1 Height of desk				
2 Side of calculator				
3 Width of door				
4 Side of classroom				

Choose your own lengths to estimate and measure.

Lengths	Metric Units		U.S. Customary Units	
	Estimate	Measurement	Estimate	Measurement
5				
6				

Math Boxes

1 Solve.

Unit
stickers

45 + ______ = 50

70 = 61 + ______

87 + ______ = 90

______ + 33 = 40

2 In my building there are 15 dogs, 13 cats, and 12 birds. How many pets are there in all?

Answer: ______ pets

3 Measure this line segment in centimeters.

About ______ cm

Draw a line segment that is 3 centimeters shorter than the one above.

What is the length of the line you drew?

About ______ cm

4 Solve. You can draw a diagram to help.

In the morning it was 60°F. At noon it was 75°F. How much did the temperature change?

Number model:

____

Answer: ______ °F

5 **Writing/Reasoning** How did you add the numbers in Problem 2? Explain your strategy.

____

____

Record of Our Arm Spans

Lesson 7-6

DATE

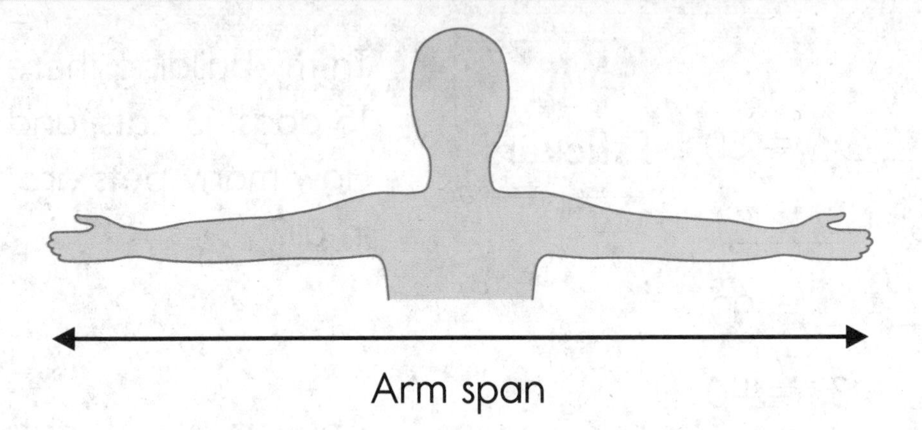

Work with a partner to measure your arm span in both inches and centimeters. Record your measurements below.

1. My arm span is about ______ inches.
2. My arm span is about ______ centimeters.

You will complete the next question in Lesson 7-8.

3. Which has a larger number:
   your measurement in inches or
   your measurement in centimeters? ________________ Why?

   ____

   ____

   ____

Record of Our Jumps

Lesson 7-6

DATE

You will make two jumps. For each jump, measure to the nearest centimeter and to the nearest inch. Here's how to measure each jump:

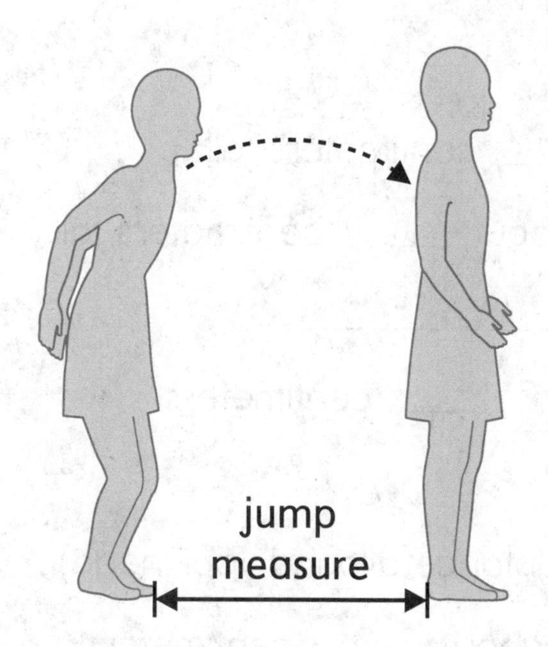

Place a penny or other marker (or make a dot with chalk) where the Jumper's back heel lands. Measure from the starting line to the marker.

1. Record your two jumps.

First jump:

About ______ centimeters

About ______ inches

Second jump:

About ______ centimeters

About ______ inches

2. Circle the measurements of your longer jump.

You will complete the next question in Lesson 7-7.

3. Which standing jump length has the most Xs above it on your line plot? ____________________

Comparing Measurements

Lesson 7-6

DATE

Work with a partner. Measure your height, head size, and shoe length to the nearest centimeter. For each measurement, choose a tool to use. You may use a ruler, a meterstick, or a tape measure.

1. Height

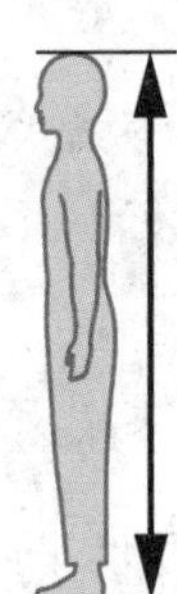

I am about ______ centimeters tall.

My partner is about ______ centimeters tall.

Who is taller? ______________

How much taller? ______ centimeters

2. Head size (the distance around your head)

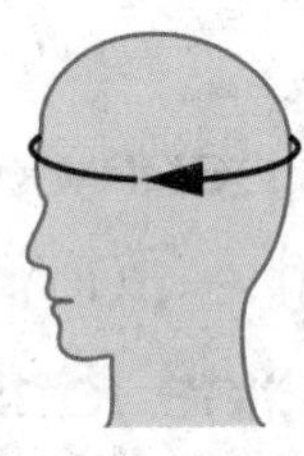

My head size is about ______ centimeters.

My partner's head size is about ______ centimeters.

Who has the larger head size? ______________

How much larger? ______ centimeters

3. Shoe length

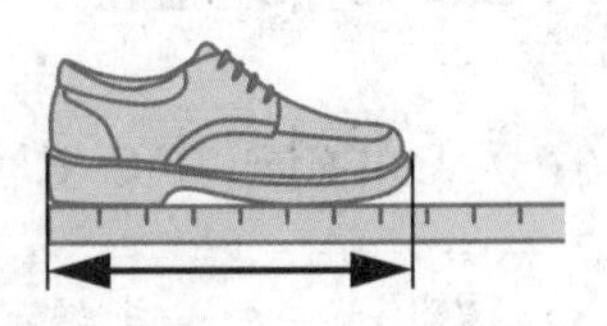

My shoe is about ______ centimeters long.

My partner's shoe is about ______ centimeters long.

Who has the longer shoe length? ______________

How much longer? ______ centimeters

Math Boxes
Preview for Unit 8

Lesson 7-6

DATE

1 Draw or write the name of something in the classroom shaped like a circle.

MRB 122–123

2 Which picture shows a cube?

Fill in the circle next to the correct answer.

 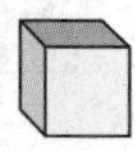

MRB 134–136

3 Count by 2s. Write the number of dots.

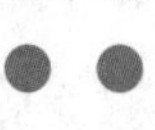

_____ dots

4 Count to find the number of squares.

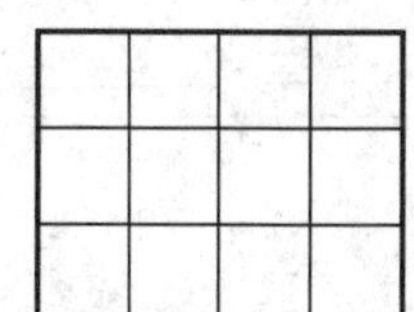

_____ squares

5 Draw a line from each number sentence to the matching array.

4 + 4 + 4 + 4 = 16   3 + 3 = 6   2 + 2 + 2 + 2 + 2 = 10

```
X X X        X X        X X X X
X X X        X X        X X X X
             X X        X X X X
             X X        X X X X
             X X
```

Solving Subtraction Problems

Lesson 7-7

DATE

Make a ballpark estimate. Then solve.

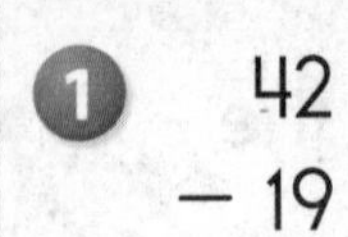

1. $\begin{array}{r} 42 \\ -\ 19 \\ \hline \end{array}$

Unit

Ballpark estimate:

2. $\begin{array}{r} 64 \\ -\ 39 \\ \hline \end{array}$

Unit

Ballpark estimate:

3. $\begin{array}{r} 86 \\ -\ 57 \\ \hline \end{array}$

Unit

Ballpark estimate:

Try This

4. $\begin{array}{r} 103 \\ -\ 58 \\ \hline \end{array}$

Unit

Ballpark estimate:

Math Boxes

1. During the picnic I counted 14 grasshoppers, 16 flies, and 25 ants. How many insects did I count in all?

   Answer: ______ insects

2. Write each number in expanded form.

   591 ____________________

   311 ____________________

   702 ____________________

   920 ____________________

   MRB 72

3. The green snake is 10 cm long. The brown snake is 26 cm long. How much longer is the brown snake than the green snake? Draw a diagram to help.

   Number model:

   ____

   Answer: ______ cm longer

4. Dan has 20 red blocks and 30 blue blocks. He gives 12 blocks to his sister. How many blocks does he have left?

   Write one or two number models:

   ____

   Dan has ______ blocks left.

5. **Writing/Reasoning** Look at Problem 3. If the green snake grows 10 cm, will it be longer than the brown snake? Explain.

   ____

   ____

   ____

Table of Our Arm Spans

Lesson 7-8

DATE

Make a table of the arm spans of your classmates.

Our Arm Spans

Arm Span (inches)	Frequency	
	Tallies	Number
	Total =	

Line Plot of Our Arm Spans

Lesson 7-8

DATE

Make a line plot of your class arm span data.

Our Arm Spans

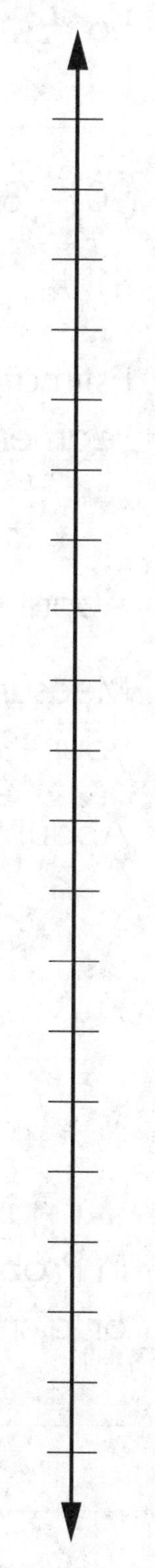

Math Boxes

DATE

1 Solve.

Unit
stickers

56 + _____ = 60

_____ + 65 = 70

90 = 82 + _____

41 + _____ = 50

2 Solve.

Unit
tacks

14 + 3 + 13 = _____

_____ = 12 + 6 + 18

5 + 14 + 18 = _____

40 = 15 + 6 + _____

3 Estimate the length of this line segment in centimeters.

About _____ centimeters

Measure the line segment with your 10-centimeter ruler.

About _____ centimeters

4 Solve. You can draw a diagram to help.

At 6:00 A.M. the temperature was 45°F. By noon it went up 15°F. What was the temperature at noon?

Number model:

____

Answer: _____ °F

5 **Writing/Reasoning** Suppose you measured the line segment in Problem 3 in inches. Would your answer in inches be a larger or a smaller number than your answer in centimeters? Explain.

____

____

Math Boxes

DATE

Math Boxes

1. The first-grade class has 23 children. The second-grade class has 21 children. The third-grade class has 17 children. How many children are there in all?

   Answer: ______ children

2. Write each number in expanded form.

   463 ____________________

   205 ____________________

   640 ____________________

   888 ____________________

   MRB 72

3. The red rope is 45 feet long. The yellow rope is 30 feet long. How much longer is the red rope than the yellow rope? You may draw a diagram to help.

   Number model:

   ____

   Answer: ______ feet

4. Emma had 30 stickers. Her teacher gave her 12 more. Then Emma gave 9 stickers to her friend. How many stickers does Emma have now?

   Number model:

   ____

   Emma has ______ stickers.

5. **Writing/Reasoning** For Problem 2, Luka wrote 40 + 60 + 3 as the expanded form of 463. Do you agree with Luka? Explain.

   ____

   ____

Sorting Shapes

Lesson 7-9

DATE

Trace 3 shapes from one of your sorts. Write a name that describes the sort.

Sorting Shapes (continued)

Lesson 7-9

DATE

Trace 3 shapes from a different sort. Write a name that describes the sort.

Drawing a Picture Graph

Lesson 7-9

DATE

Tally chart:

Name of Fruit	Number of Children
Apple	
Banana	
Grapes	
Other	

Picture graph:

Our Favorite Fruits

Name of Fruit

KEY: ☺ = 1 child

Measuring Body Parts

Lesson 7-9

DATE

Complete the chart.

Body Part	Centimeters	Inches
Wrist	Estimate: About ______ centimeters Measurement: About ______ centimeters	Estimate: About ______ inches Measurement: About ______ inches
Arm span	Estimate: About ______ centimeters Measurement: About ______ centimeters	Estimate: About ______ inches Measurement: About ______ inches
Around head	Estimate: About ______ centimeters Measurement: About ______ centimeters	Estimate: About ______ inches Measurement: About ______ inches
Waist	Estimate: About ______ centimeters Measurement: About ______ centimeters	Estimate: About ______ inches Measurement: About ______ inches
Height	Estimate: About ______ centimeters Measurement: About ______ centimeters	Estimate: About ______ inches Measurement: About ______ inches

Math Boxes Preview for Unit 8

1. Draw or write the name of something in the classroom that is shaped like a rectangle.

2. Look at a centimeter cube. Write one thing you notice about it.

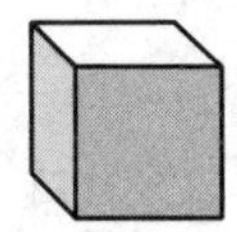

____

____

3. Count by 5s. Write the number of dots.

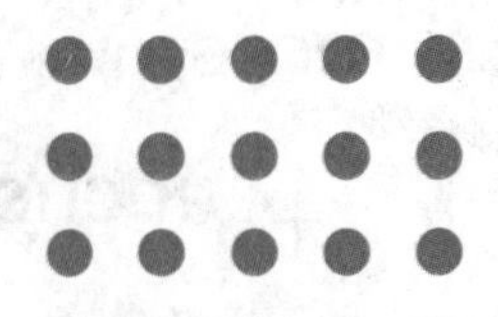

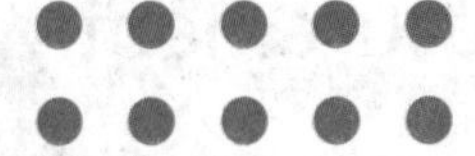

______ dots

4. Count to find the number of squares.

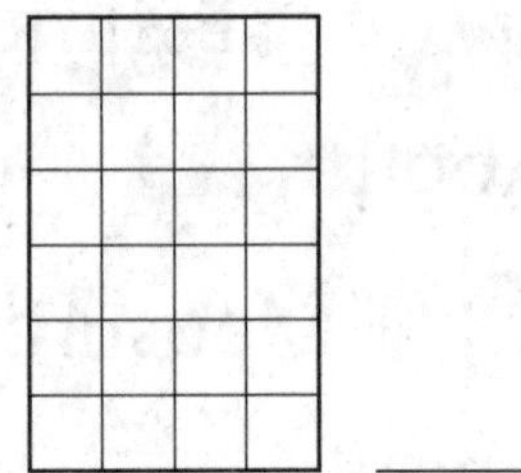

______ squares

5. Draw a line from each number sentence to the matching array.

$2 + 2 + 2 + 2 = 8$　　$5 + 5 + 5 + 5 + 5 = 25$　　$4 + 4 + 4 = 12$

X X X X X
X X X X X
X X X X X
X X X X X
X X X X X

X X X X
X X X X
X X X X

X X
X X
X X
X X

Math Boxes

Lesson 8-1

DATE

1 Solve.

Unit

$46 + 21 + 9 + 34 =$ ______

2 The den window is 42 cm wide. The kitchen window is 60 cm wide. How much wider is the kitchen window than the den window?

Answer: ______ cm wider

3 Write a number model. Solve. You may draw a diagram.

The Spartans scored 76 points. The Huskies scored 40 points. How many more points did the Spartans score?

Number model:

____

Answer: ______ points

MRB 30–31

4 Use Xs to show the following length data on the line plot.

Lengths of Objects (inches): 6, 5, 9, 8, 6, 7, 8, 9, 7, 9, 7

Lengths of Objects

5 6 7 8 9 10

Length of Object (inches)

5 **Writing/Reasoning** Explain your strategy for solving Problem 1.

____

____

____

Math Boxes

1 Solve.

Unit

49 + ______ = 80

100 = 65 + ______

42 + ______ = 90

______ = 60 − 23

2 Solve.

Unit

35 + 11 + 15 + 29 = ______

3 Measure this line segment to the nearest centimeter:

About ______ cm

Draw a line segment that is 4 centimeters shorter:

How long is the line segment you drew? About ______ cm

4 Sarah has 35 shells. Dana has 35 shells. Together, they gave away 30 shells. How many shells do they have left? Circle the letter of the correct answer.

A 40 shells **B** 50 shells

C 70 shells **D** 100 shells

MRB 34–35

5 **Writing/Reasoning** Write a number story to match this number model: 24 − 12 = ?

______

______

Solve your number story. Answer: ______ (unit)

Comparing Pentagons and Hexagons

Lesson 8-3

DATE

Use straws and twist ties.

1. Build a 5-sided shape.
   Draw the shape below.

   Shape name: ______________

2. Build a different 5-sided shape.
   Draw the shape below.

   Shape name: ______________

3. How are the shapes you drew for Problems 1 and 2 alike?
   How are they different?

   ____

   ____

4. Build a 6-sided shape.
   Draw the shape below.

   Shape name: ______________

5. Build a different 6-sided shape.
   Draw the shape below.

   Shape name: ______________

6. How are the shapes you drew for Problems 4 and 5 alike?
   How are they different?

   ____

   ____

Math Boxes

1 Solve.

Unit

34 + 20 + 16 + 25 = ______

2 The white flagpole is 38 feet tall. The gray flagpole is 20 feet tall. How much taller is the white flagpole than the gray one?

Answer: ______ feet taller

3 Solve. You may draw a diagram.

Christina jumped 40 inches. This is 13 more inches than Justin jumped. How many inches did Justin jump?

Number model:

____

Answer: ______ inches

4 Use Xs to show the following height data on the line plot.

Heights of Plants (centimeters): 38, 35, 36, 40, 37, 38, 36, 38, 40

Heights of Plants

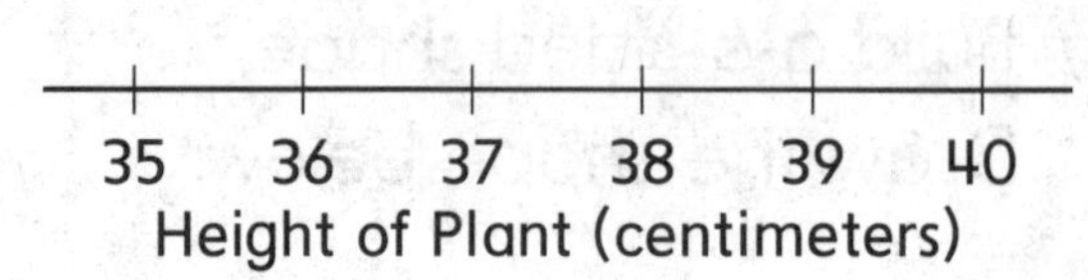

5 **Writing/Reasoning** Dan wrote this number model for Problem 3: 13 + ? = 40. Do you agree? Explain your answer.

____

____

____

MRB 30–31

DATE

What Is a Quadrilateral?

These shapes are quadrilaterals:

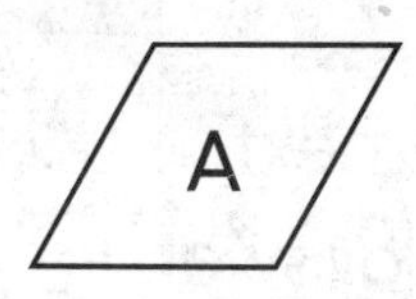

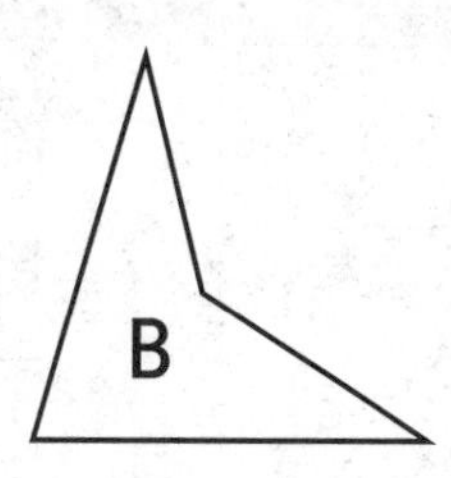

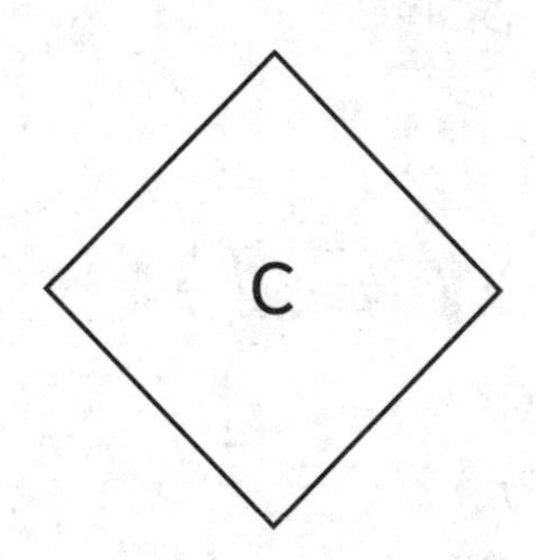

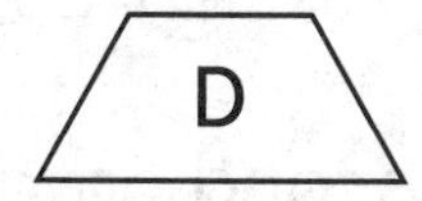

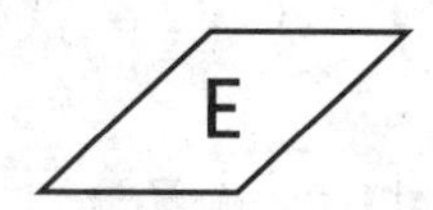

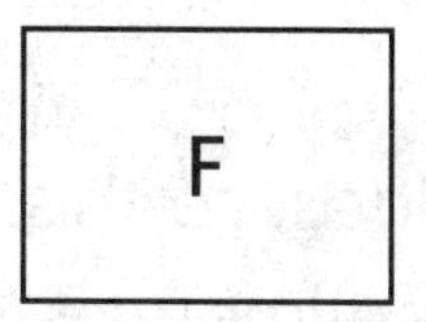

These shapes are not quadrilaterals:

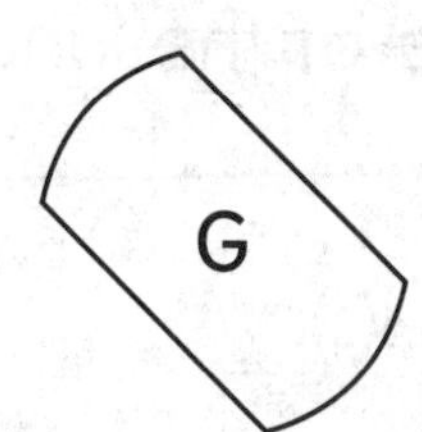

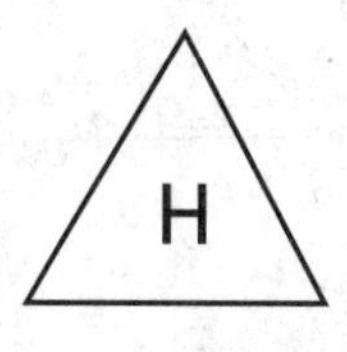

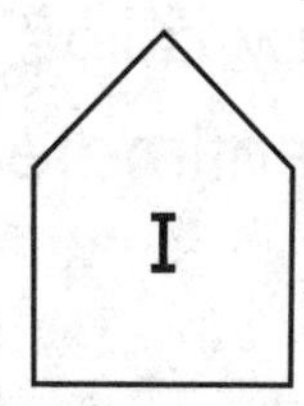

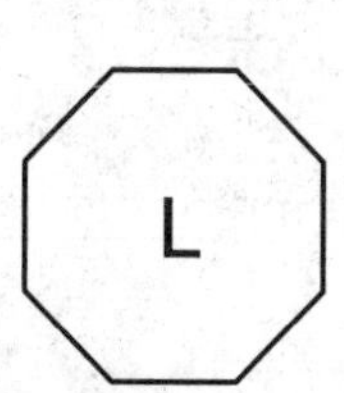

Look at the shapes.

What do the quadrilaterals have in common?

What do you think are the attributes of a quadrilateral?

Tell your partner your ideas.

Math Boxes

1 Solve.

Unit

21 + ______ = 50

100 = 52 + ______

______ = 70 − 19

38 + ______ = 90

2 Solve.

Unit

Fill in the circle next to the correct answer.

23 + 18 + 12 + 19 = ______

Ⓐ 62 Ⓑ 72

Ⓒ 73 Ⓓ 83

3 Measure this line segment in centimeters.

About ______ cm

Draw a line segment 4 centimeters longer.

How long is the line segment you drew?

About ______ cm

MRB 104

4 There are 15 second graders at the lunch table. Then 12 more sit down. After a while, 10 children leave. How many are at the lunch table now?

Number model(s):

____

There are ______ children.

5 **Writing/Reasoning** Write a number story to match the number model: 35 + 15 = ?

____

____

Solve your number story.

Answer: ______________________ (Unit)

Math Boxes

Lesson 8-5

DATE

1 Fill in the unit box.
Solve.

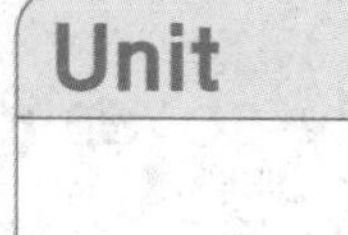

37 + ______ = 100

______ = 29 + 71

______ + 42 = 100

68 + 32 = ______

2 Estimate the width of your *Math Journal*:

About ______ inches

Measure it to the nearest inch:

About ______ inches

MRB 101–102

3 About how long is a large swimming pool? Circle the closest measurement.

50 centimeters

50 inches

50 meters

4 Solve. Show your work.

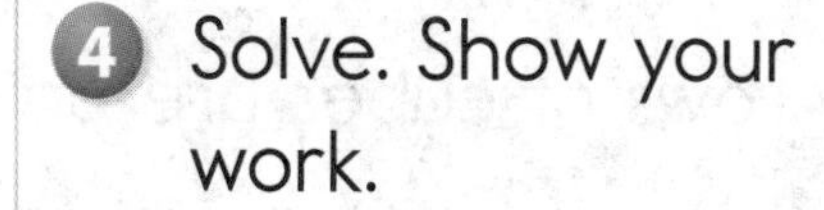

Unit

95 − 41 = ______

5 **Writing/Reasoning** Look at Problem 2. Suppose you measured the width of your *Math Journal* in centimeters. Would the number of centimeters be more than or less than the number of inches? Explain.

____

____

____

Covering Rectangles

Lesson 8-6

DATE

Math Message

1. Use centimeter cubes to completely cover Rectangle A without gaps or overlaps. Then answer the questions.

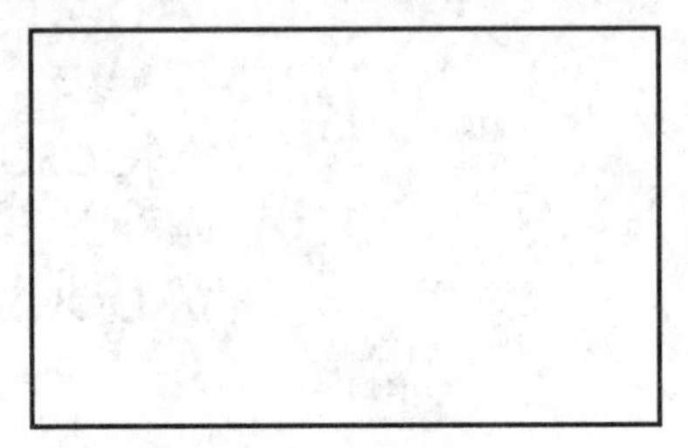

Rectangle A

How many rows of cubes are on Rectangle A? ______

How many cubes are there in each row on Rectangle A? ______

How many cubes did you use to cover Rectangle A? ______

2. Draw squares on Rectangle B to show how you covered Rectangle A with cubes.

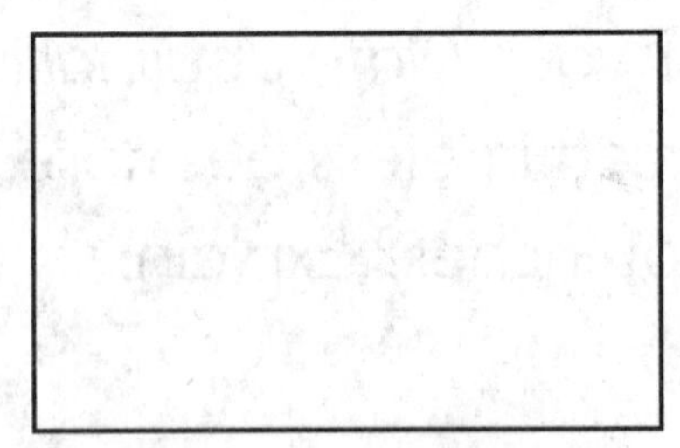

Rectangle B

Covering Rectangles (continued)

3 • Take one square pattern block.

• Use it to figure out how many square pattern blocks can completely cover Rectangle C with no gaps or overlaps.

• Draw lines to show how the squares cover the rectangle.

Rectangle C

How many rows of squares are there? ______

How many squares are in each row? ______

How many squares are there in all? ______

Partitioning Rectangles

Lesson 8-6

DATE

Partition each rectangle into squares that are the same size as a square pattern block. You may use a pattern block to help you.

1

How many rows? ______ How many squares in each row? ______

How many squares are there in all? ______

2

How many rows? ______ How many squares in each row? ______

How many squares are there in all? ______

Math Boxes

DATE

Unit

Math Boxes

1. Circle the letters next to all the pictures that show quadrilaterals.

 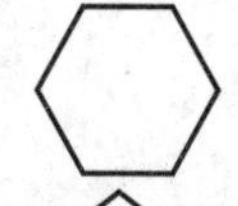

C

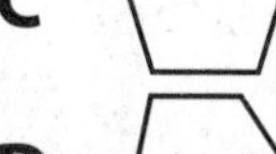

2. 68 + 34 = ?
Use partial-sums addition to solve.

Ballpark estimate:

____

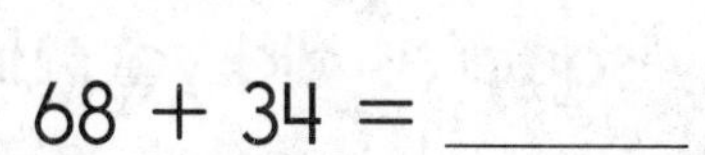

68 + 34 = ______

3. Write each number in expanded form.

302 ______________

476 ______________

862 ______________

590 ______________

4. Draw a shape that has 4 sides.

5. Solve. Show your work on an open number line.

Brooklyn has 25 yellow blocks and 42 green blocks. How many blocks does she have in all? ______ blocks

Partitioning Strategies

Lesson 8-7

DATE

Math Message

1. Use one square pattern block to partition the rectangle without any gaps or overlaps. Draw where you placed your square each time.

   How many rows of squares? ______

   How many squares in each row? ______

   How many squares did you draw to cover the whole rectangle? ______

Partition each rectangle below into same-size squares. Use the small square to help you. Draw squares to show your partitioning.

2. How many rows of squares are there? ______

   How many squares are in each row? ______

   How many squares did you draw to cover the rectangle? ______

3. How many rows of squares are there? ______

   How many squares are in each row? ______

   How many squares did you draw to cover the rectangle? ______

Partitioning into Same-Size Squares

Lesson 8-7

DATE

1. Partition this square into 2 rows with 2 same-size small squares in each row.

   How many small squares cover the large square? ______

2. Partition this rectangle into 5 rows with 7 same-size squares in each row.

   How many squares cover the rectangle? ______

Solving Addition Problems

Lesson 8-7

DATE

Fill in the unit box. Then solve each problem. You can draw an open number line or use a number line, number grid, or base-10 blocks to help you. Show your work.

Unit

1. $\begin{array}{r} 38 \\ +\ 41 \\ \hline \end{array}$

2. $\begin{array}{r} 26 \\ +\ 57 \\ \hline \end{array}$

3. $\begin{array}{r} 454 \\ +\ 365 \\ \hline \end{array}$

4. $\begin{array}{r} 258 \\ +\ 667 \\ \hline \end{array}$

Math Boxes

Lesson 8-7

DATE

Math Boxes

1 Fill in the unit box. Solve.

Unit

53 + 47 = ______

18 + ______ = 100

______ + 76 = 100

______ = 69 + 31

2 Estimate how long your pencil is in centimeters:

About ______ centimeters

Measure it to the nearest centimeter:

About ______ centimeters

3 About how long is a school playground? Circle the closest measurement.

100 inches

100 centimeters

100 yards

4 Solve. Show your work.

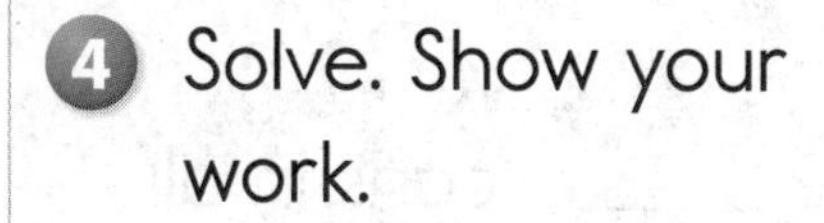

Unit

95 − 31 = ______

5 **Writing/Reasoning** Look at Problem 1. For ______ + 76 = 100, Emily said the missing number is 34. Do you agree? Explain.

____

____

____

____

MRB
76–78, 80

Equal-Groups and Array Number Stories

Lesson 8-8

DATE

For each number story, draw a picture. Then answer the question and write a number model.

1. There are 3 rows of cans on the shelf with 4 cans in each row. How many cans are there in all?

   There are ______ cans in all.

   Number model:

   ____

2. Mr. Yung has 4 boxes of markers. There are 6 markers in each box. How many markers does he have in all?

   Mr. Yung has ______ markers.

   Number model:

   ____

3. Sandy has 4 bags of marbles. Each bag has 4 marbles. How many marbles does she have in all?

   Sandy has ______ marbles.

   Number model:

   ____

Try This

4. Mei folded her paper into 2 columns of 4 boxes each. How many boxes did she make?

   Mei made ______ boxes.

   Number model:

   ____

Math Boxes
Preview for Unit 9

Lesson 8-8

DATE

1. Divide the square into 2 equal parts.

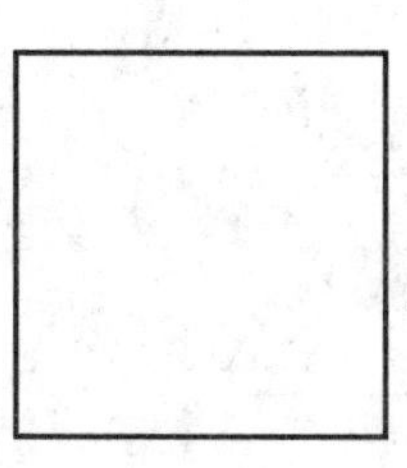

2. Solve. Show your work.

$$\begin{array}{r} 82 \\ -\ 29 \\ \hline \end{array}$$

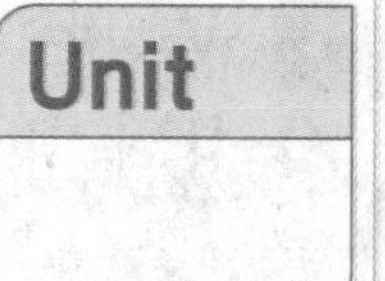

3. Which number shows 3 in the ones place, 5 in the hundreds place, and 0 in the tens place? Circle the correct answer.

A 350

B 305

C 530

D 503

4. Count by fives. How many dots in all?

Answer: ______ dots

Write a number model.

____

5. Make a ballpark estimate. Write a number model to show your estimate.

149¢ + 38¢ = ?

My estimate:

____

6. Measure the line segment to the nearest inch.

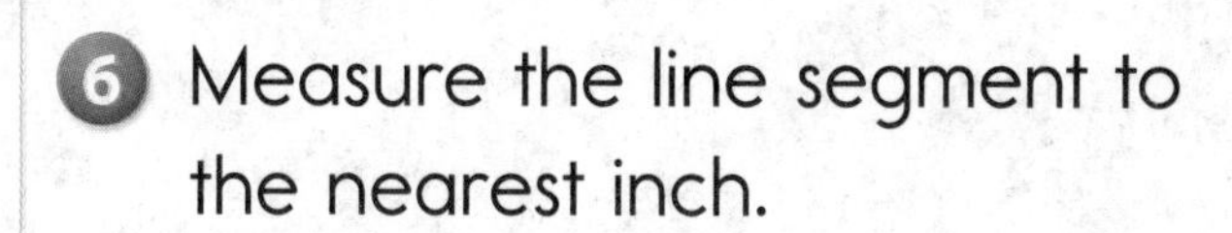

About ______ inches

Equal Groups and Arrays

Lesson 8-9

DATE

Math Message

1. Write a number story to match the number model $4 + 4 + 4 = 12$. Draw a picture of your story.

____

____

____

2. Draw the last set of equal groups you or your partner made from counters.

3. Draw the last array you or your partner made from counters.

4. Write a number model for the equal groups and the array you drew.

____

Math Boxes

Lesson 8-9

DATE

1 How many more points did Team A score than Team C?

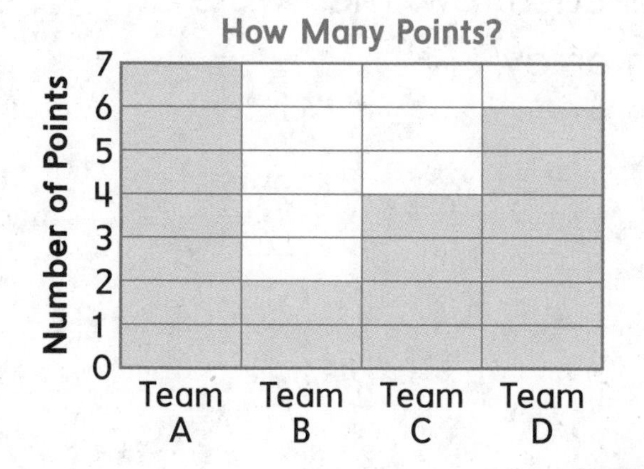

Answer: ______ points

MRB 116

2

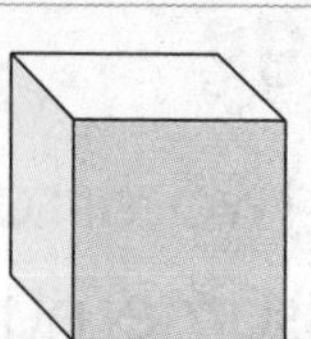

What shape is each face?

____

MRB 122–123, 134–135

3 Solve.

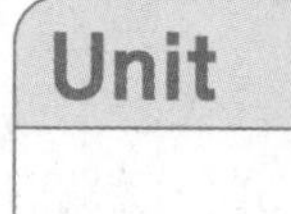

15 + 11 + 45 + 19 = ______

4 Solve using partial-sums addition.

$$\begin{array}{r} 112 \\ +\ 42 \\ \hline \end{array}$$

Ballpark estimate:

____

MRB 78–80

5 **Writing/Reasoning** Explain how you solved Problem 3.

____

____

____

Arranging Desks

Lesson 8-10

DATE

Math Message

Draw at least two arrays to show how you can put 12 desks in rows with the same number of desks in each row. Then write an addition number model to match each array.

Solving Subtraction Problems

Lesson 8-10

DATE

Fill in the unit box. For each problem, make a ballpark estimate and solve. You can use a number line, a number grid, or base-10 blocks to help you.

Unit

1

$$\begin{array}{r} 36 \\ -\ 25 \\ \hline \end{array}$$

Ballpark estimate:

____

2

$$\begin{array}{r} 50 \\ -\ 28 \\ \hline \end{array}$$

Ballpark estimate:

____

3

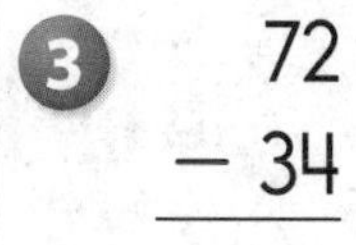

$$\begin{array}{r} 72 \\ -\ 34 \\ \hline \end{array}$$

Ballpark estimate:

____

Try This

4

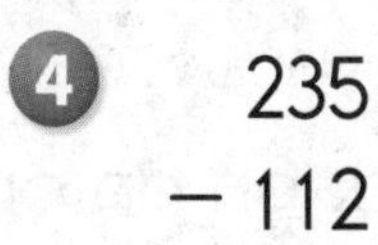

$$\begin{array}{r} 235 \\ -\ 112 \\ \hline \end{array}$$

Ballpark estimate:

____

Math Boxes

Lesson 8-10

DATE

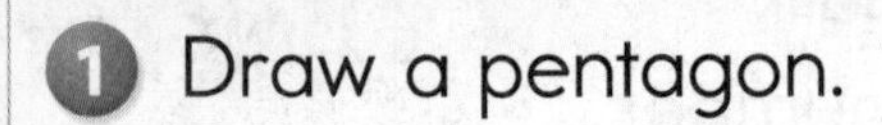

1. Draw a pentagon.

MRB 123

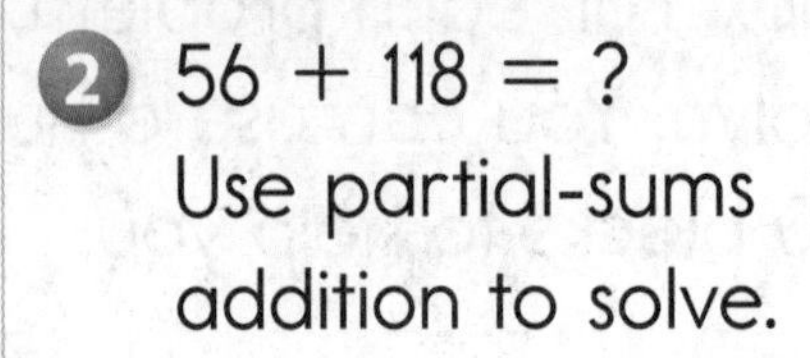

2. 56 + 118 = ?
Use partial-sums addition to solve.

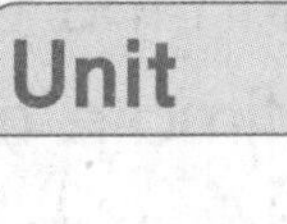

Ballpark estimate:

____

56 + 118 = ______

3. Write each number in expanded form.

706 ______________________

418 ______________________

880 ______________________

749 ______________________

MRB 73

4. Circle the hexagon.

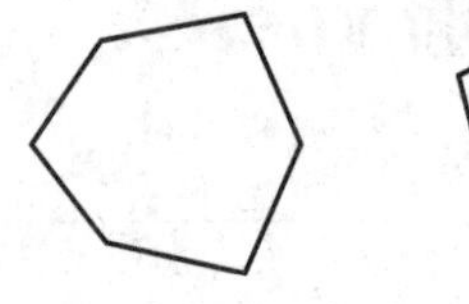

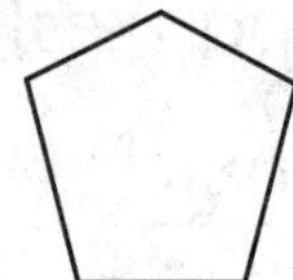

MRB 123

5. Solve. Show your work on an open number line.

Jean has a box of 48 crayons. Joe has a box of 24 crayons. How many crayons do they have all together?

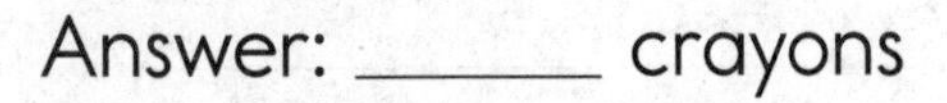

Answer: ______ crayons

Mystery Shapes

Lesson 8-11

DATE

Draw the mystery shapes in the space below. Write the name of each shape under your drawing.

_______________ _______________

Math Boxes

Lesson 8-11

DATE

Math Boxes

1 How many more pine trees are there than maple trees?

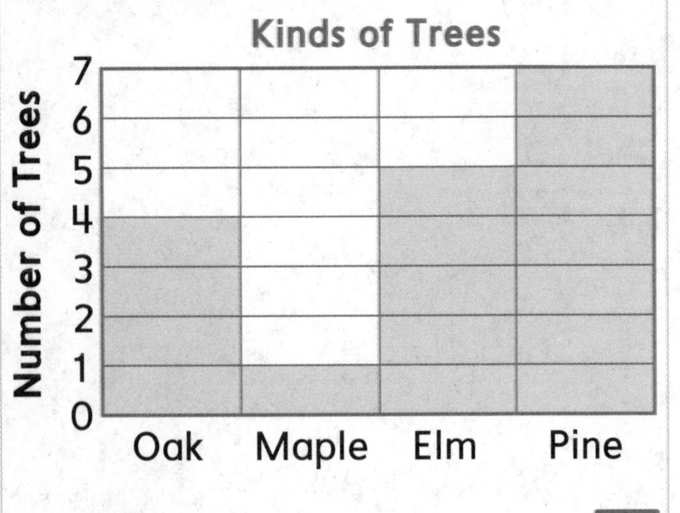

Answer: ______ trees

MRB 116

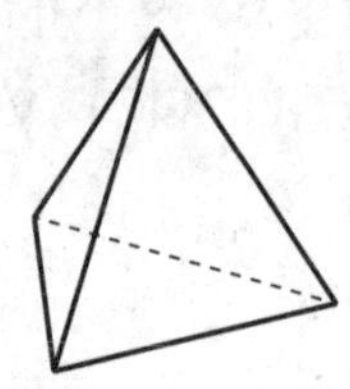

What shape is each face?

____

MRB 122–123, 134–135

3 Solve.

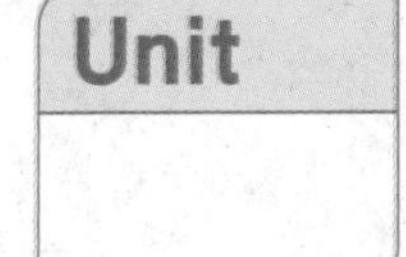

$9 + 41 + 20 + 18 =$ ______

4 Solve using partial-sums addition.

$$\begin{array}{r} 131 \\ +\ 57 \\ \hline \end{array}$$

Ballpark estimate:

____

MRB 79–80

5 **Writing/Reasoning** In Problem 1, Natalie says there are 17 trees in all. Allison says there are 27 trees in all. Who is correct? Explain. ______________________

____

____

Math Boxes
Preview for Unit 9

Lesson 8-12

DATE

Math Boxes

1. Divide the rectangle into 4 equal parts.

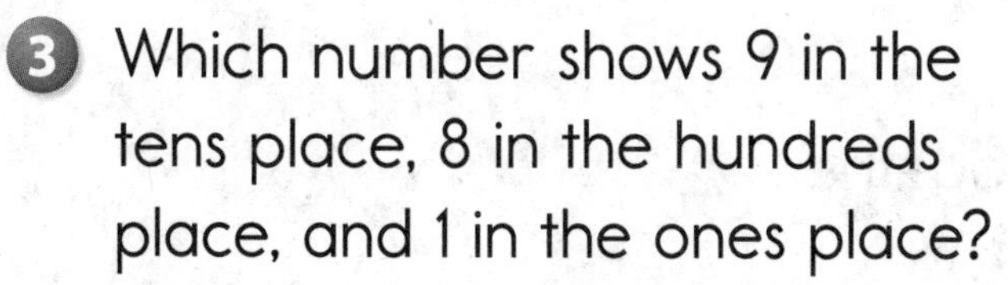

MRB 132–133

2. Solve. Show your work.

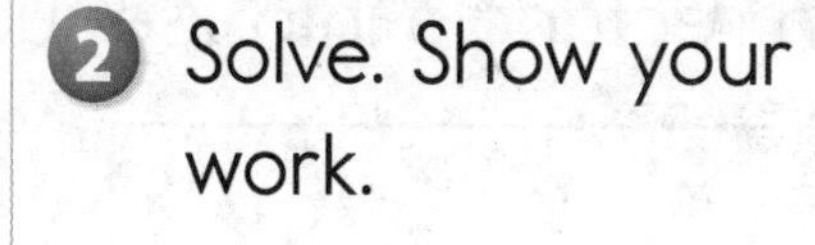

Unit

$$\begin{array}{r} 98 \\ -\ 75 \\ \hline \end{array}$$

MRB 81–89

3. Which number shows 9 in the tens place, 8 in the hundreds place, and 1 in the ones place?

Fill in the circle next to the correct answer.

Ⓐ 891 Ⓑ 198

Ⓒ 819 Ⓓ 918

MRB 73

4. ● ● ● ● ● ●
● ● ● ● ● ●

Count by 2s. How many dots in all?

Answer: _____ dots

Write a number model.

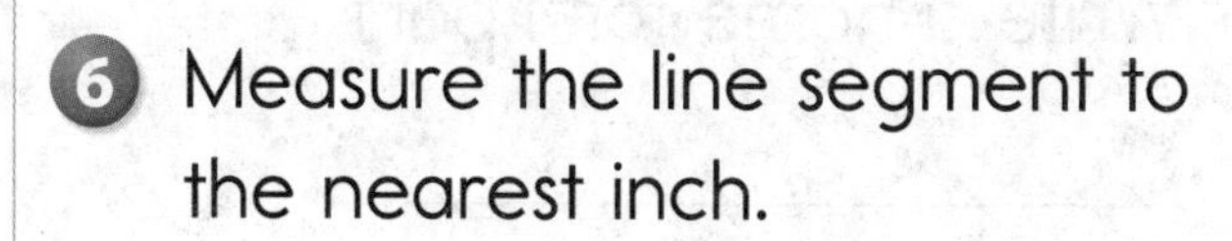

____

____

5. Make a ballpark estimate. Write a number model to show your estimate.

49¢ + 129¢ = ?

My estimate:

_______________ ¢

MRB 79

6. Measure the line segment to the nearest inch.

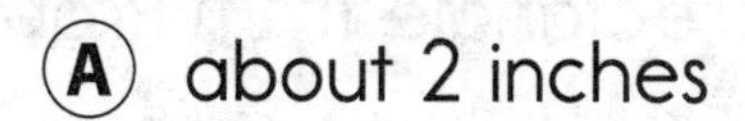

Fill in the circle next to the correct answer.

Ⓐ about 2 inches

Ⓑ about 1 inch

Ⓒ about 4 inches

MRB 101

Showing Equal Parts

Lesson 9-1

DATE

1. Divide each rectangle into 2 equal parts. Show two different ways.

Write a name for 1 part.

____

Write a name for all of the parts together.

____

2. Divide each rectangle into 3 equal parts. Show two different ways.

Write a name for 1 part.

____

Write a name for all of the parts together.

____

3. Pick one rectangle from Problem 2. How could you show that the parts are equal? ____________________________

____

____

Showing Equal Parts (continued)

Lesson 9-1

DATE

4. Divide this rectangle into 4 equal parts.

Write a name for 1 part.

____

Write a name for all of the parts together.

____

5. Ruthie divided this rectangle into 3 parts. She named one of the parts "one-third."

Do you agree with Ruthie? ______
Explain your answer.

____

____

Try This

6. Divide this circle into 4 equal parts.

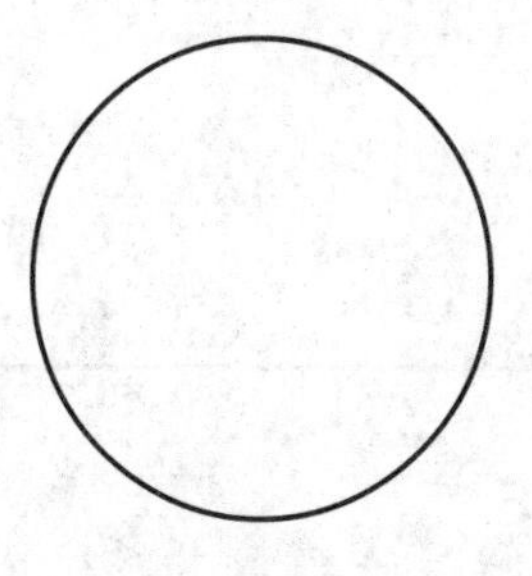

1 Draw an X on the shape that has more angles.

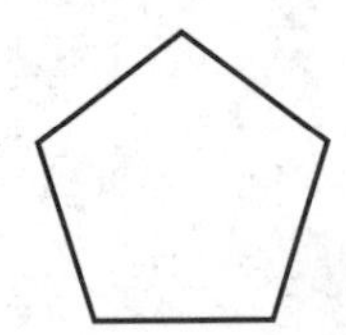
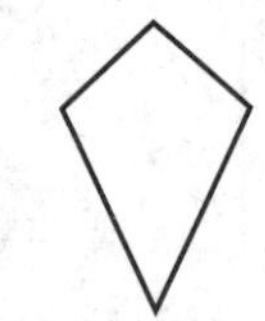

Write a name for the shape you drew an X on.

____

2 Write the number in expanded form.

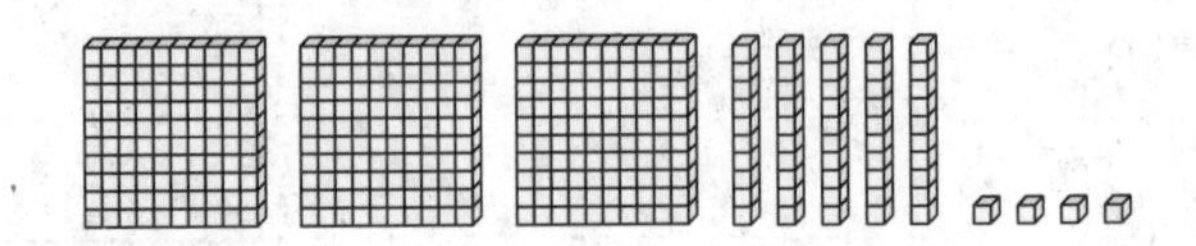

____

3 Draw same-size squares to completely cover the rectangle. Use the small square on the left to help you.

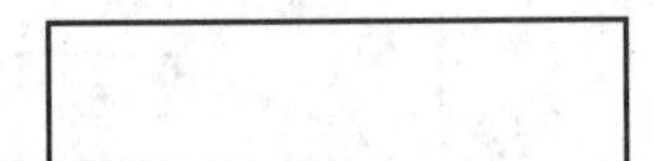

How many squares cover the rectangle? ______

4 For each sentence, circle the unit that makes sense.

Ted's arm is 2 ______ long.

feet inches

Mo's goldfish is 8 ______ long.

centimeters meters

The building is 15 ______ tall.

inches yards

5 Solve this problem. Use an open number line to show your work.

$60 - 35 =$ ______

MRB
84–85

Sharing a Cracker

Lesson 9-2

DATE

Math Message

Juan has a cracker he wants to share equally with 2 friends.

He divided the cracker in an unusual way:

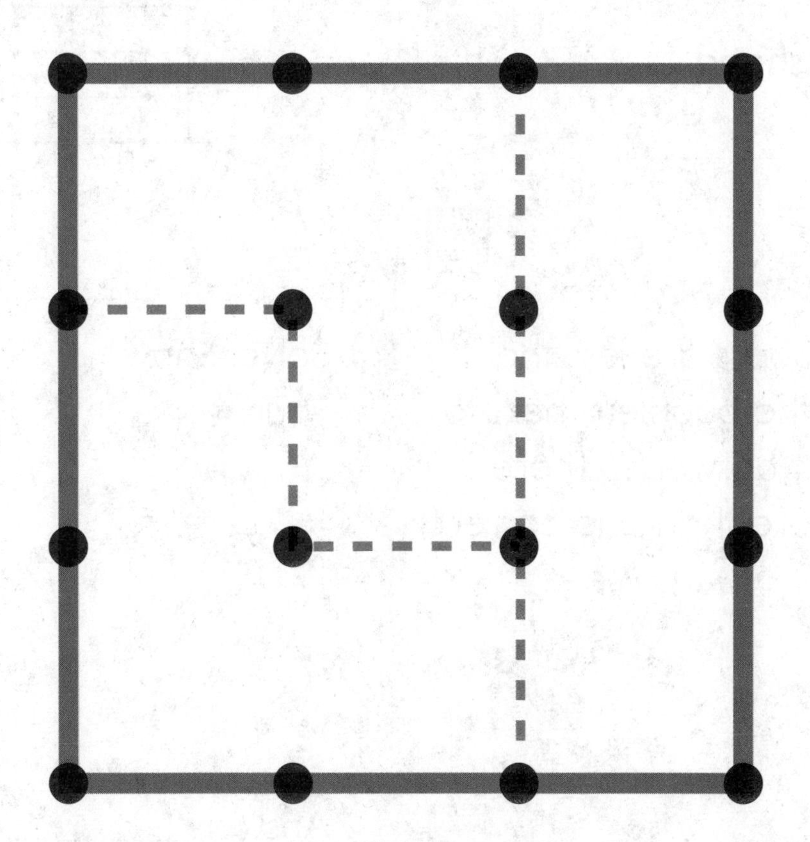

Do you think Juan divided the cracker into 3 equal parts? ______

Explain your answer.

Math Boxes

Lesson 9-2

DATE

Math Boxes

1 Draw a shape that has 5 sides.

Name this shape.

____

2 Divide this shape into 2 equal parts.

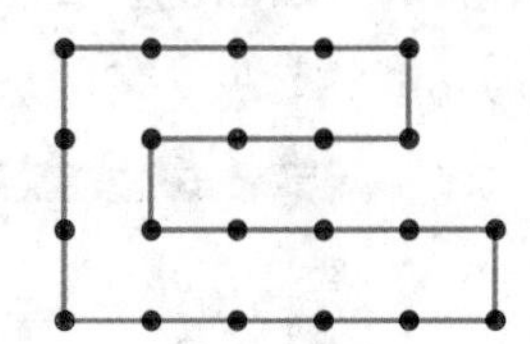

3 Which numbers are less than 601? Fill in the bubble(s) next to the correct answer(s). There may be more than one correct answer.

- ◯ 599
- ◯ 711
- ◯ 803
- ◯ 487

4 Solve. Show your work.

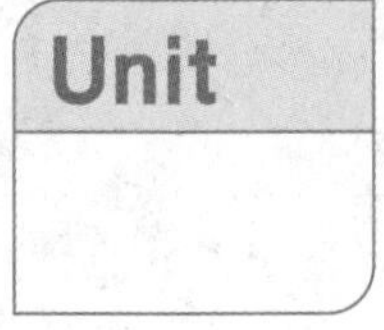

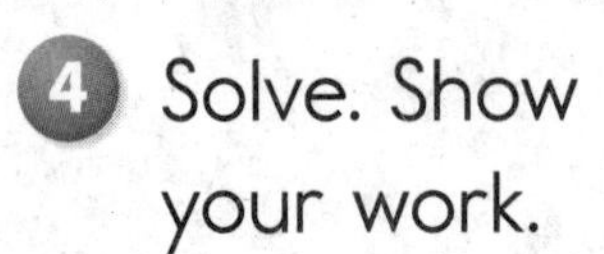

$$\begin{array}{r} 94 \\ -\ 37 \\ \hline \end{array}$$

Answer: ______

5 **Writing/Reasoning** In Problem 1 you drew a shape with 5 sides. Draw a shape with 5 sides that looks different from that one.

What is the name of your shape? ____________ How do you know?

____

An Art Project

Lesson 9-3

DATE

A teacher is preparing to cut a piece of construction paper into three equal shares to use in an art project. The teacher drew dotted lines to guide the cuts.

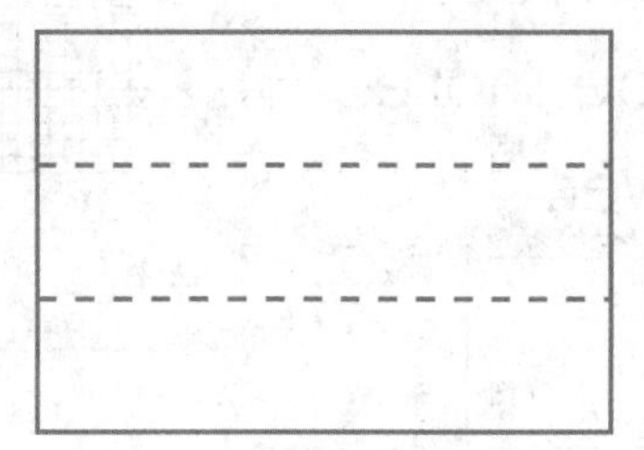

Two children are talking about the paper. Jaylan says, "The paper is three-thirds." Leila says, "No, it's not. It's one whole."

Who is correct? Talk with a partner about your ideas.

Math Boxes

DATE

Math Boxes

1. Which shapes are quadrilaterals? Fill in the bubble(s) next to the correct answer(s). There may be more than one correct answer.

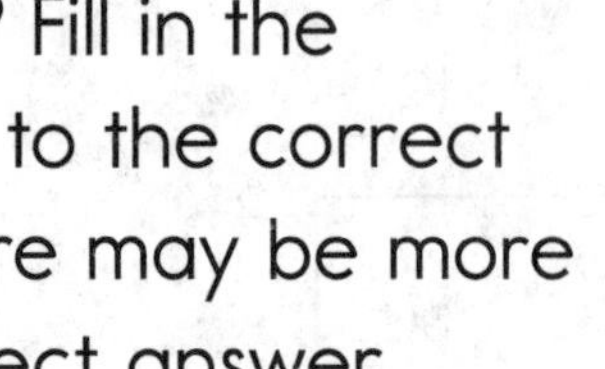

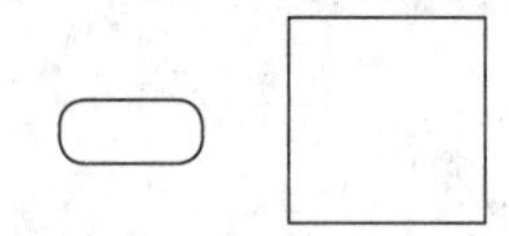

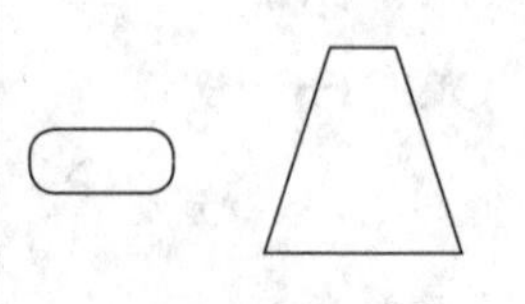

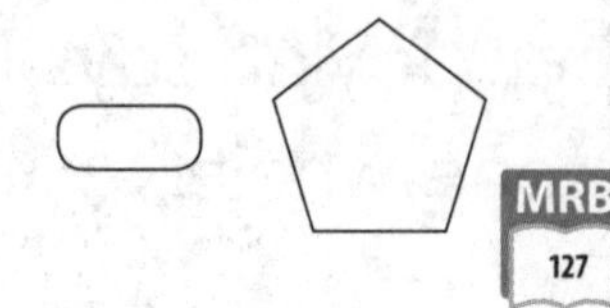

MRB 127

2. Write the number in expanded form.

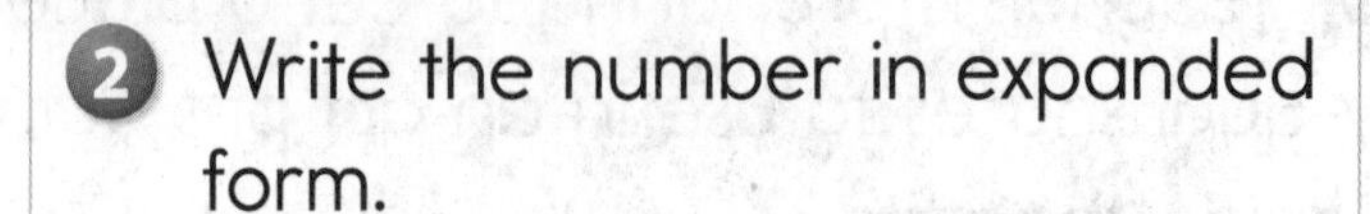

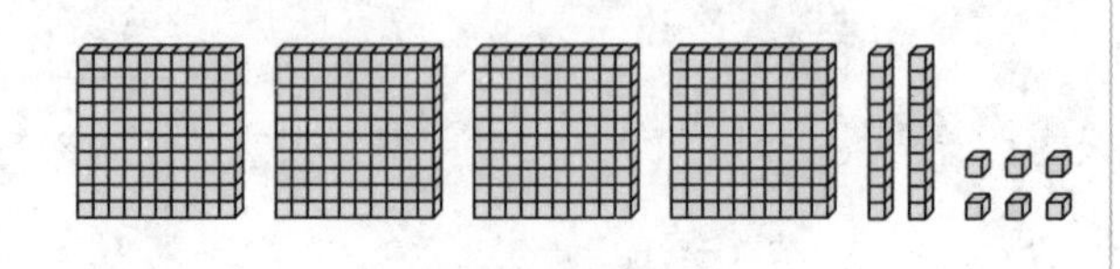

____

3. Draw same-size squares to completely cover the rectangle. Use the small square to help.

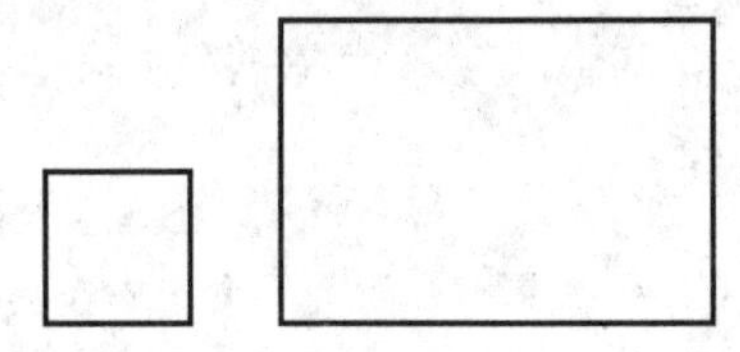

I drew ______ squares.

4. For each sentence, circle the unit that makes sense.

A football field is
100 ______ long.
yards   inches

Jessi's bedroom is
4 ______ long.
centimeters   meters

5. Solve the problem. Use an open number line to show your work.

85 − 34 = ______

Measuring Lengths

Lesson 9-4

DATE

Use your 12-inch ruler to measure the lengths of these objects to the nearest inch.

1 Crayon

About ______ inches long

2 Eraser

About ______ inches long

Use your 12-inch ruler to measure the lengths of these objects to the nearest half-inch.

3 Marker

About ____________________ inches long

4 Paintbrush

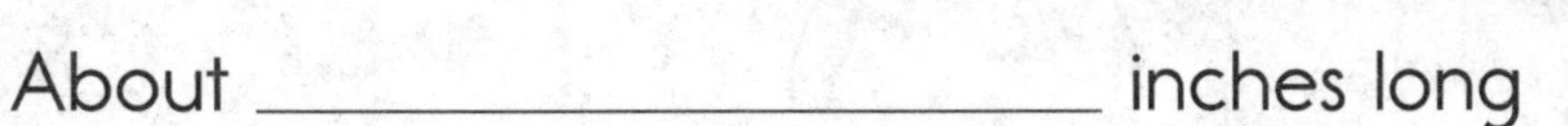

About ____________________ inches long

Measure the length of your desk to the nearest half-inch.

5 My desk is about ____________________ inches long.

Use the measurements above to complete Problem 6.

6 The crayon is about ______ inches longer than the eraser.

Partitioning Circles into Halves, Thirds, and Fourths

Lesson 9-4

DATE

1 Divide this circle into 2 equal parts.

Write a name for 1 part.

____

Write a name for all of the parts together.

____

2 Divide the circle into 4 equal parts.

Write a name for 1 part.

____

Write a name for all of the parts together.

____

3 Which circle is divided into thirds (or 3 equal parts)? ______________

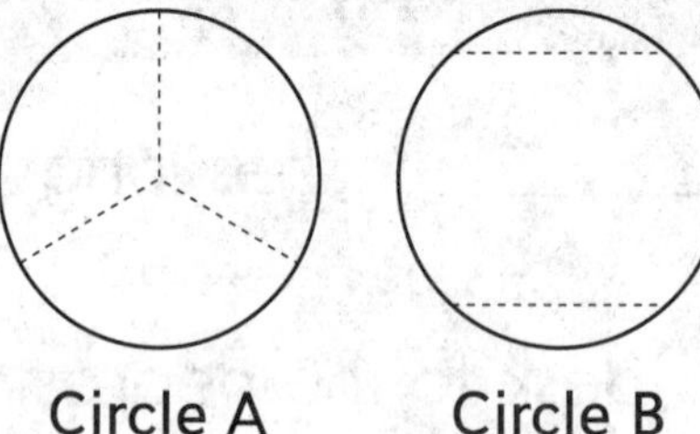

How do you know? ____

____

____

For the circle divided into thirds, write a name for all of the parts.

____

Math Boxes

1. Write the name of this shape.

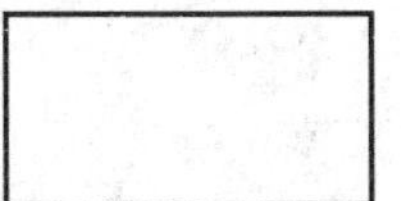

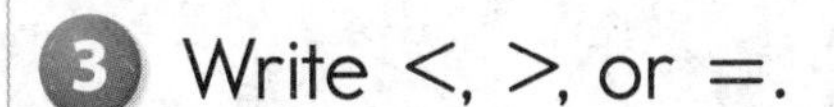

MRB 127–128

2. Divide this shape into 2 equal parts.

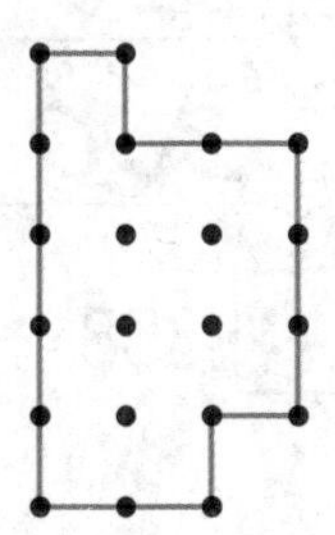

3. Write <, >, or =.

341 ______ 431

202 ______ 199

511 ______ 511

989 ______ 899

MRB 74–75

4. Solve. Show your work.

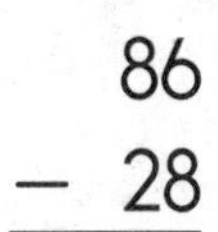

$$\begin{array}{r} 86 \\ -\ 28 \\ \hline \end{array}$$

Unit

Answer: ______

5. **Writing/Reasoning** Compare the shape in Problem 1 to a square. How are they alike? How are they different?

Comparing Numbers

DATE

Write each number in expanded form. Then write < or > in the box to compare the two numbers.

Example: 325 = 300 + 20 + 5

358 = 300 + 50 + 8

325 [<] 358

1. 136 = ______________

   224 = ______________

   136 [ ] 224

2. 765 = ______________

   756 = ______________

   765 [ ] 756

Write < or > in the box.

3. 503 [ ] 530

4. 147 [ ] 417

Try This

Write each number in expanded form. Then write < or > in the box to compare the two numbers.

5. 1,583 = ______________

   1,771 = ______________

   1,583 [ ] 1,771

6. 2,944 = ______________

   2,908 = ______________

   2,944 [ ] 2,908

Math Boxes

Lesson 9-5

DATE

1 Divide the circle into 2 equal parts.

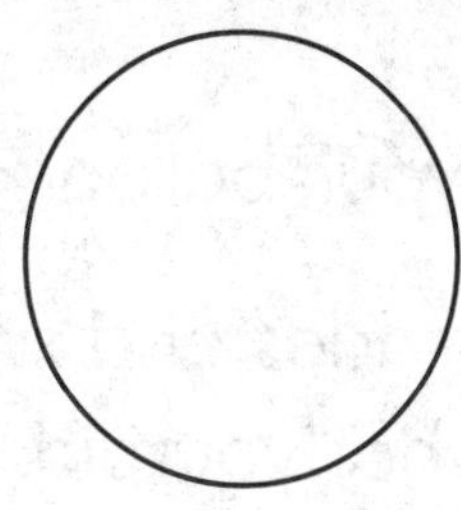

Use words to name 1 part.

____

Use words to name both parts.

____

MRB 132

2

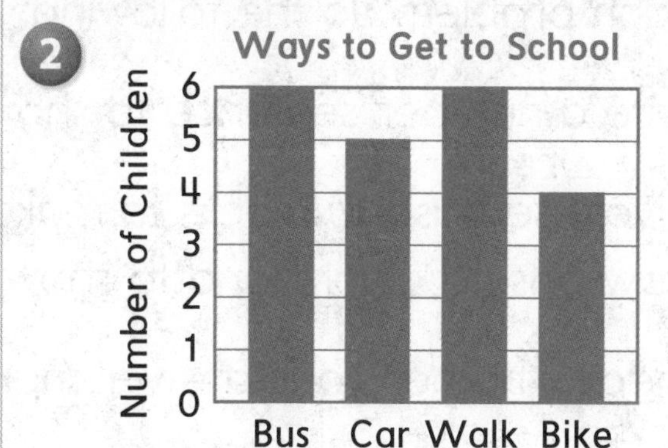

How many children in all walk or bike to school? ______

3 What is the value of the digit 4 in each number?

14 ______

142 ______

436 ______

4 Write the number word for 612.

____

5 Which number models match the array? Fill in the bubbles next to all the number models that match.

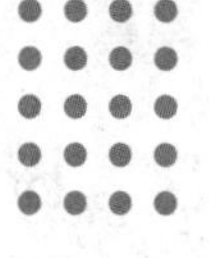

- ⬭ $5 + 5 = 10$
- ⬭ $4 + 4 + 4 + 4 + 4 = 20$
- ⬭ $4 + 4 = 8$
- ⬭ $5 + 5 + 5 + 5 = 20$

6 Frank shared a sandwich with a friend. He cut it like this:

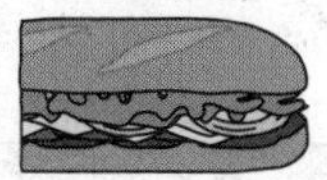

What part of the sandwich does each child get?

____

Subtracting with Base-10 Blocks

Lesson 9-6

DATE

Unit

For each problem do the following:

- Write a number sentence to show your ballpark estimate.
- Solve. Use base-10 blocks to make trades and subtract. Draw base-10 shorthand to show what you did.
- Check whether your answer makes sense.

1 68 − 17 = ? Ballpark estimate: ______________ Solution: Answer: ______	**2** 81 − 64 = ? Ballpark estimate: ______________ Solution: Answer: ______
3 95 − 36 = ? Ballpark estimate: ______________ Solution: Answer: ______	**4** 144 − 127 = ? Ballpark estimate: ______________ Solution: Answer: ______

Math Boxes

Lesson 9-6

DATE

Math Boxes

1 Divide this shape into 3 equal parts.

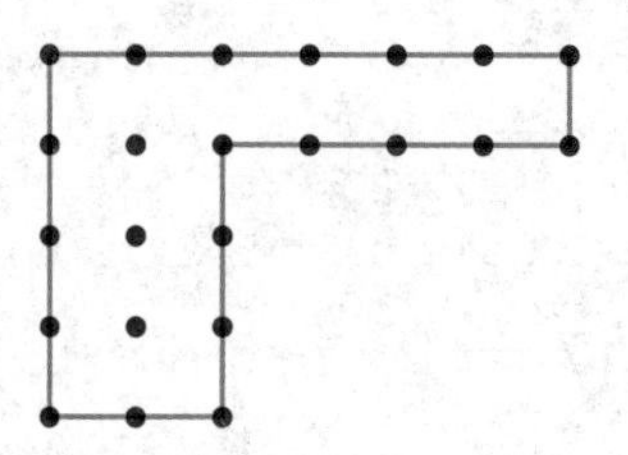

2 How many dots are in this array?

Write a number model to solve.

____

3 Four children will share 3 crackers. Use the rectangles below to show how they can share the crackers equally.

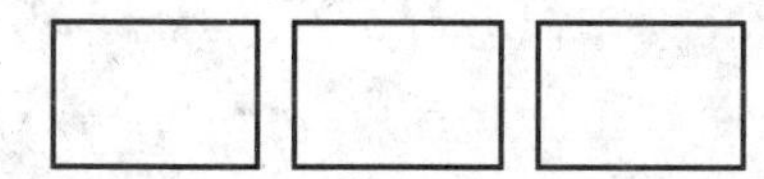

4 How many small squares will it take to cover the rectangle? Draw the squares.

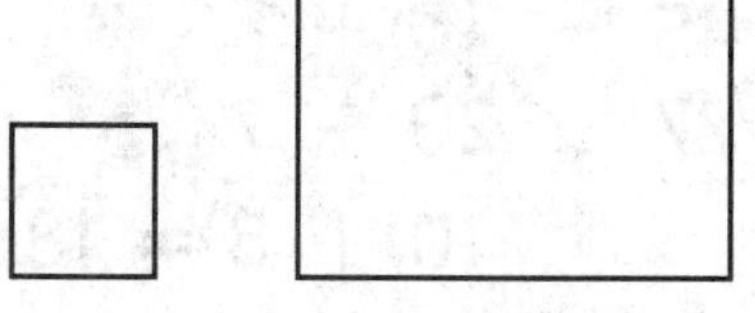

_______ squares

5 **Writing/Reasoning** In Problem 3, how much is in each child's share?

____

____

____

Expand-and-Trade Subtraction

Lesson 9-7

DATE

For each problem do the following:

- Write a number model to show a ballpark estimate.
- Write each number in expanded form.
- Solve using expand-and-trade subtraction.

1 Example: 45 − 27 = ?

Ballpark estimate:

45 - 20 = 25

Solution:

$$\begin{array}{r} 45 \\ -\ 27 \\ \hline \end{array} \quad \begin{array}{l} \rightarrow \overset{30}{\cancel{40}} + \overset{15}{\cancel{5}} \\ \rightarrow 20 + 7 \\ \hline 10 + 8 = 18 \end{array}$$

45 - 27 = 18

2 31 − 17 = ?

Ballpark estimate:

____

Solution:

31 − 17 = ______

Expand-and-Trade Subtraction (continued)

Lesson 9-7

DATE

3 $72 - 49 = ?$

Ballpark estimate: ______________

Solution:

$72 - 49 =$ ______

Try This

$126 - 48 = ?$

Ballpark estimate: ______________

Solution:

$126 - 48 =$ ______

Math Boxes

Lesson 9-7

DATE

Math Boxes

1 Divide the circle into 4 equal parts. Use words to name 1 part.

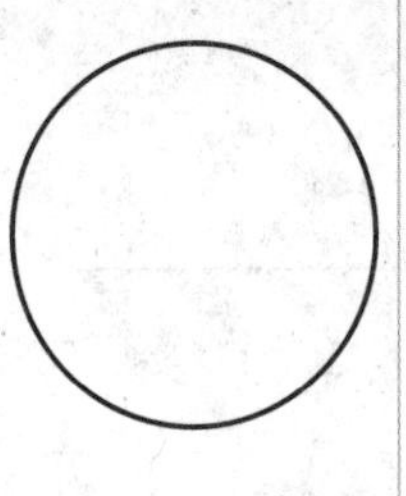

____

Use words to name all the parts.

____

MRB 132–133

2

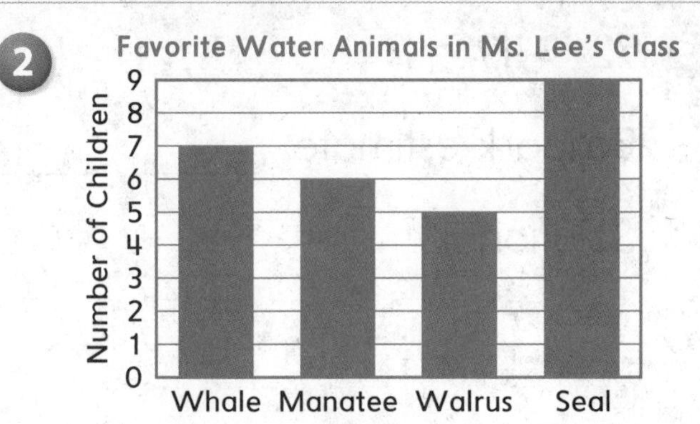

How many children in all like whales or manatees best?

_____

3 What is the value of the digit 6 in each number?

26 _____

679 _____

160 _____

4 Write the number word for 840.

____

5 • • • • • • • •
• • • • • • • •

How many dots are there in all? _____

Write a number model for the array. _____________

6 Rosita and 3 friends equally shared an orange. How much does each child get?

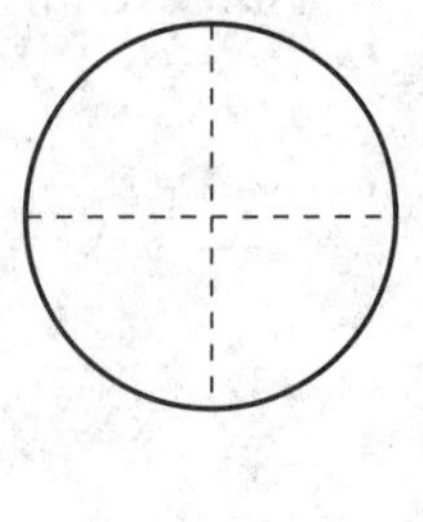

- One-third
- Two-fourths
- Four-fourths
- One-fourth

Math Boxes

Lesson 9-8

DATE

1. Divide this shape into 3 equal parts.

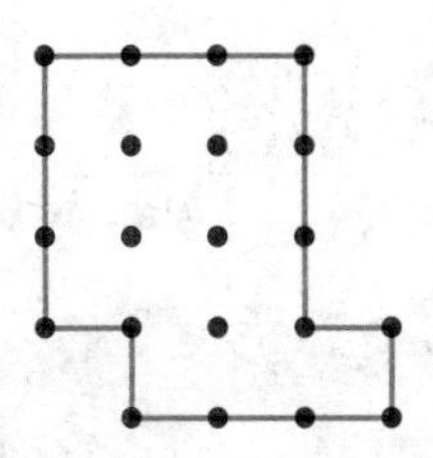

2. How many dots are there in all?

Write a number model to solve.

____

3. Six children share 4 muffins. Use the circles below to show how they can share the muffins equally.

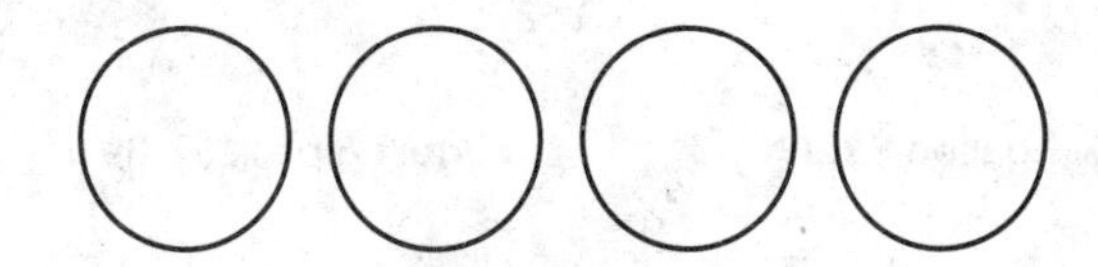

MRB 132–133

4. How many small squares will it take to cover the large square? Draw the squares.

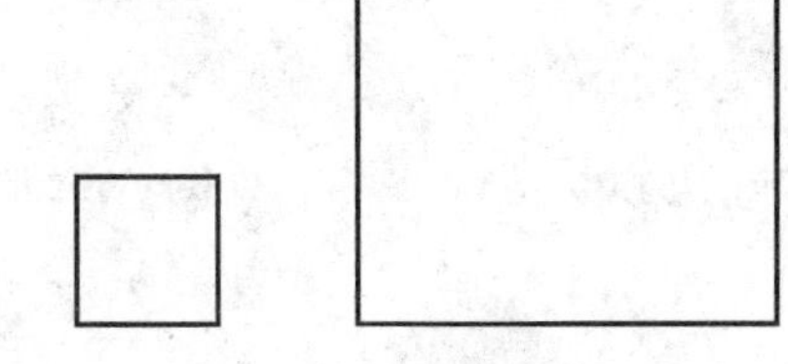

_____ squares

5. **Writing/Reasoning** In Problem 2, Amy wrote $3 + 6 = 9$ as a number model for the array. Is she correct? Explain.

____

____

____

Good Buys Poster

DATE

Fruit Group	Meat/Protein Group
Plums 69¢ lb **Oranges $1.49** lb **Bananas 59¢** lb **Watermelons $2.99** each	**Beans** 16 oz can **2/89¢** **Almond Butter** 8 oz jar **$1.29** **Ground Beef $3.40** lb **Chunk Light Tuna** 6.5 oz can **69¢** **Lunch Meat** 1 lb package **$1.39**
Vegetables Group	**Milk Group**
Carrots 1 lb bag **3/$1.00** **Celery 59¢** lb **Tomatoes 99¢** lb	**Milk** 1 gallon **$2.39** **Yogurt** 6 pack **$2.09** **American Cheese** 8 oz package **$1.49**
Grain Group	**Other Items**
Gluten-Free Bread 16 oz loaf **99¢** **Saltine Crackers** 1 lb box **69¢** **Hamburger Buns** 16 oz package **69¢**	**Mayonnaise** 32 oz jar **$1.99** **Ketchup** 14 oz bottle **$1.09** **Grape Jelly** 22 oz jar **$1.69**

Ways to Pay

Lesson 9-8

DATE

Math Boxes

Do the following for each item:

- Write the cost of the item.
- Find two ways to pay for it with exact change.
  Use Ⓠ, Ⓓ, Ⓝ, Ⓟ, and [$1] to record your answers.
  Example: You buy 1 lb of bananas. They cost 59¢ for a pound.
  You pay with Ⓠ Ⓠ Ⓝ Ⓟ Ⓟ Ⓟ Ⓟ or Ⓓ Ⓓ Ⓓ Ⓓ Ⓓ Ⓝ Ⓟ Ⓟ Ⓟ Ⓟ.

In Problems 3–4, choose your own items to buy.

1. You buy a can of chunk light tuna.

   Cost: ____________

   Pay with | **or**

2. You buy a loaf of gluten-free bread.

   Cost: ____________

   Pay with | **or**

3. You buy ____________________

   ____________________.

   Cost: ____________

   Pay with | **or**

4. You buy ____________________

   ____________________.

   Cost: ____________

   Pay with | **or**

Making Estimates

Lesson 9-9

DATE

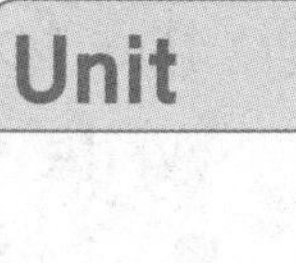

Jayden makes a ballpark estimate for the problem below. He uses numbers that are close to the numbers in the problem, but easier to use.

$48 + 24 = ?$

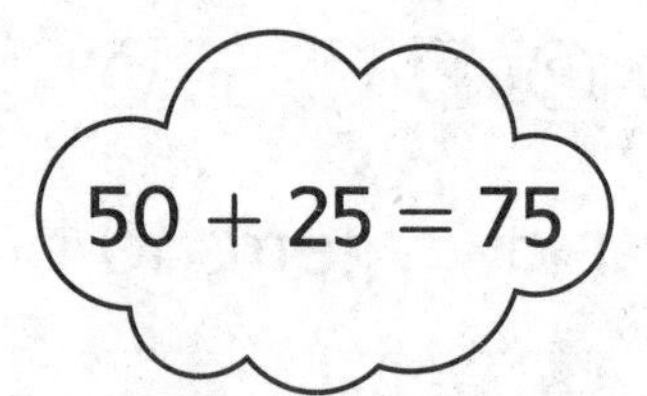

Jayden's estimate

1. Explain Jayden's thinking to a partner.
2. Make a different ballpark estimate for Jayden's problem. What close-but-easier numbers could you use?

Math Boxes

1. Which is NOT the value of the digit 3 in the number 368?

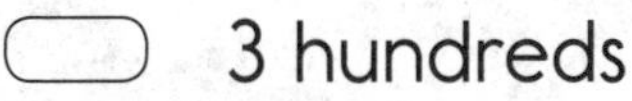

- ◯ 3 hundreds
- ◯ 30 tens
- ◯ 300 ones
- ◯ 3 ones

2. Write <, >, or =.

598 ______ 592

157 ______ 257

1,925 ______ 1,297

3. Make a ballpark estimate. Then solve.

Unit

$$\begin{array}{r} 108 \\ -\ 56 \\ \hline \end{array}$$

Ballpark estimate:

____

4. Divide the rectangle into thirds.

Use words to name 1 part.

____

Use words to name all the parts. ______________

5. Solve. Show your work on the open number line.

Lucas is 32 inches tall. His sister Martina is 59 inches tall. How much taller is Martina than Lucas? ______ inches

Two Equal Groups

Lesson 9-10

DATE

Solve the number stories. You can use cubes or draw pictures to help. Then write an addition number model for each problem. The number model should be a doubles fact.

1 There are 2 rows of paintings hanging on a wall. Each row has 3 paintings. How many paintings are on the wall?

Answer: ______ paintings

Addition number model:

____

2 A bookcase has 2 shelves. Each shelf has 9 books on it. How many books are in the bookcase?

Answer: ______ books

Addition number model:

____

3 You have 4 flowers in all and 2 vases. Each vase should have the same number of flowers. How many flowers should you put in each vase?

Answer: ______ flowers

Addition number model:

____

4 A total of 12 chairs are in 2 equal rows. How many chairs are in each row?

Answer: ______ chairs

Addition number model:

____

Try This

5 Write a multiplication number model for Problem 1.

____

6 Write a multiplication number model for Problem 2.

____

Equal Shares with Different Shapes

Lesson 9-10

DATE

Divide each shape into equal parts. Use the dots to help you draw the lines. Do not use any diagonal lines.

1. Divide this shape into 3 equal parts.

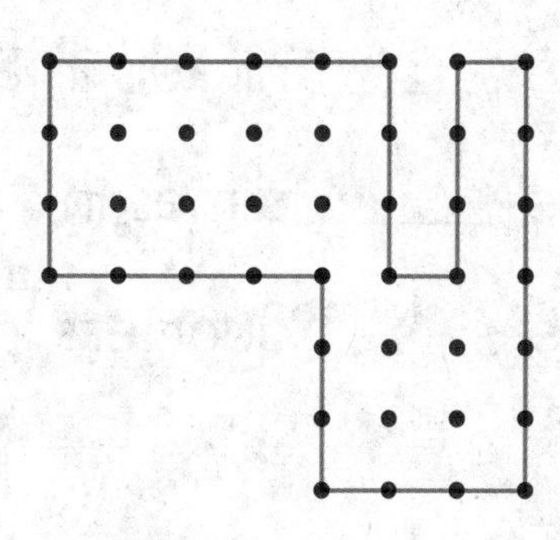

2. Divide this shape into 2 equal parts.

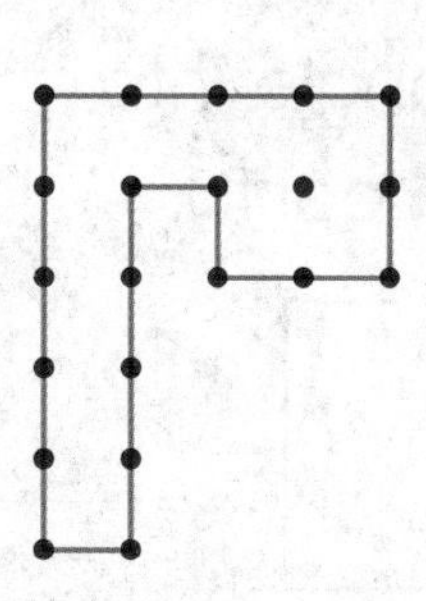

3. For Problem 2, are the parts the same shape? ______

   Are the parts the same size? ______

   How do you know?

   ____

   ____

Try This

4. Divide this shape into 4 equal parts.

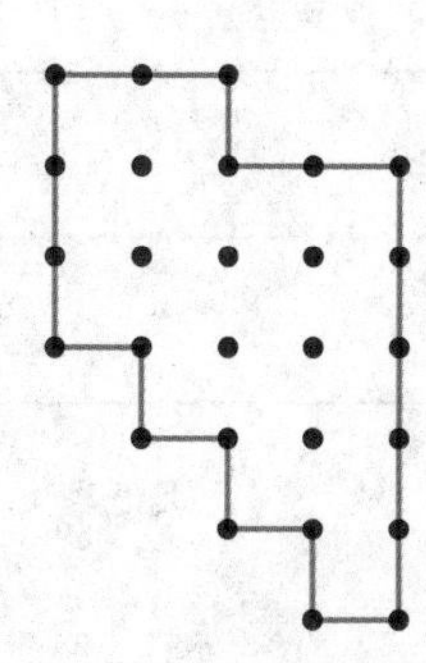

Math Boxes

Lesson 9-10

DATE

1. There are ______ pennies in $1.

   There are ______ nickels in $1.

   There are ______ dimes in $1.

   There are ______ quarters in $1.

   MRB 110–111

2. You buy juice for 109¢, or $1.09. Show 2 ways to pay with exact change. Use Ⓠ, Ⓓ, Ⓝ, and Ⓟ.

   MRB 110–111

3. Divide the rectangle into fourths (4 equal parts).

   Use words to name 1 part.

   ____

   Use words to name all the parts.

   ____

   MRB 132–133

4. Draw a quadrilateral with at least 1 pair of parallel sides.

   MRB 127–128

5. **Writing/Reasoning** In Problem 3, Joe says that one way to name all the parts together is "a whole." Is he correct? Explain.

   ____

   ____

   ____

   MRB 132–133

Dimes and Nickels

Lesson 9-11

DATE

1. Find the value of each set of dimes and the value of each set of nickels.

Number of Dimes	Value	Number of Nickels	Value
1	10 cents	1	5 cents
2	______ cents	2	______ cents
4	______ cents	4	______ cents
5	______ cents	5	______ cents
8	______ cents	8	______ cents
10	______ cents	10	______ cents

2. Look at the value of the dimes and the value of the nickels in each row. What do you notice?

____

____

Multiples of 10, 5, and 2

Lesson 9-11

DATE

1. How many is 3 [5s]? ______

   Number model:

   $3 \times 5 =$ ______

2. How many is 7 [10s]? ______

   Number model:

   $7 \times 10 =$ ______

3. How much money is 5 nickels?

   ______ cents

   Number model:

   $5 \times 5 =$ ______

4. How much money is 6 dimes?

   ______ cents

   Number model:

   $6 \times 10 =$ ______

5. How much money is 6 nickels?

   ______ cents

   Number model:

   $6 \times 5 =$ ______

6. A store display has 4 rows of oranges. There are 5 oranges in each row. Draw an array to show the oranges.

   How many oranges are in the display? ______ oranges

   How many is 4 [5s]? ______

   Number model: $4 \times 5 =$ ______

7. Jo puts pictures in an album. There are 3 rows of pictures on a page. There are 2 pictures in each row. Draw an array to show the pictures on one page.

   How many pictures are on the page? ______ pictures

   How many is 3 [2s]? ______

   Number model: $3 \times 2 =$ ______

Partitioning a Rectangle

Lesson 9-11

DATE

Paul's teacher asked him to partition the rectangle at right into 3 rows and 5 columns. Then she asked him to count the total number of small squares.

Paul's partitioned rectangle is shown at right.

How many small squares cover the rectangle? 24

1. Do you agree with Paul's solution? ______ Why or why not?

____
____
____
____

2. Show how you would partition the rectangle into 3 rows and 5 columns.

3. Describe how you partitioned the rectangle

____
____
____
____

4. How many small squares cover the rectangle?

______ squares

DATE

1. What is the value of the digit 7 in the number 1,572?

_____

MRB 73

2. Write <, >, or =.

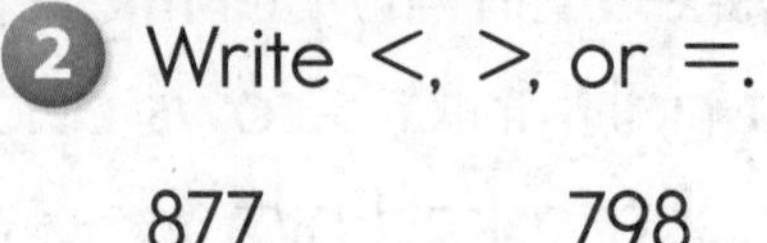

877 _____ 798

345 _____ 345

6,814 _____ 4,186

MRB 74–75

3. Make a ballpark estimate. Then solve.

Unit

$$\begin{array}{r} 345 \\ +\ 78 \\ \hline \end{array}$$

Ballpark estimate:

_____

4. Divide this square into fourths (four equal parts).

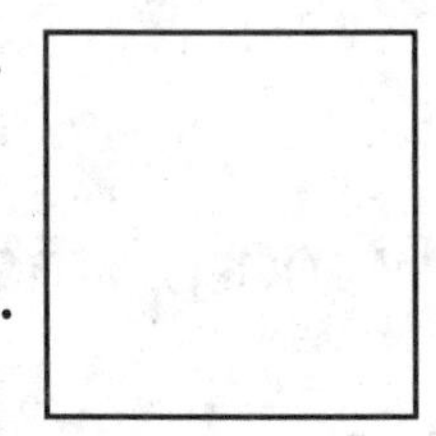

Use words to name 1 part.

_____

Use words to name all 4 parts.

_____

5. **Writing/Reasoning** Solve this number story. Show your work on the open number line.

The floor lamp is 165 cm tall. The table lamp is 70 cm tall. How much taller is the floor lamp than the table lamp? _____ cm

Math Boxes

Lesson 9-12

DATE

Math Boxes

1 _____ pennies = $2

_____ nickels = $2

_____ dimes = $2

_____ quarters = $2

2 You buy crackers for 149¢, or $1.49. Show 2 ways to pay. Use Ⓠ Ⓓ Ⓝ Ⓟ.

3 Divide the rectangle into three equal parts.

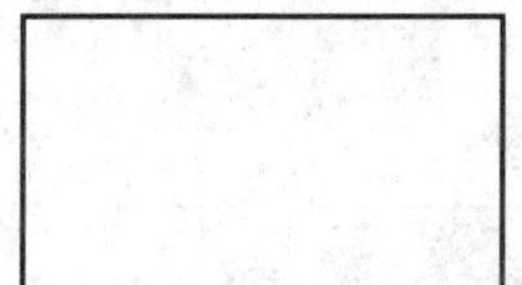

Use words to name 1 part.

_____

Use words to name all the parts.

_____

4 Circle the shape that has parallel sides.

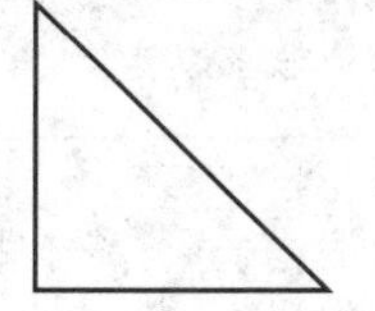

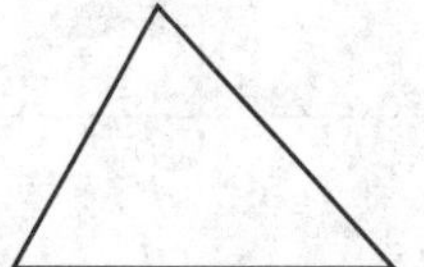

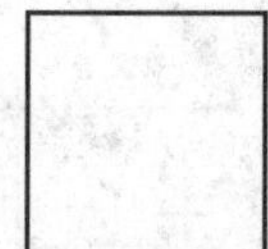

5 **Writing/Reasoning**

Explain how you know which two sides are parallel in Problem 4.

_____

_____

Addition Facts Inventory Record, Part 1

Lesson 5-9

DATE

Addition Fact	Know It	Don't Know It	How I Can Figure It Out
10 + 3			
4 + 6			
7 + 7			
3 + 2			
3 + 4			
10 + 2			
9 + 9			
2 + 2			
4 + 10			
6 + 6			
10 + 5			
5 + 2			
3 + 7			
4 + 4			

Addition Facts Inventory Record, Part 1 (continued)

Lesson 5-9

DATE

Addition Fact	Know It	Don't Know It	How I Can Figure It Out
10 + 6			
5 + 5			
8 + 10			
2 + 4			
3 + 3			
10 + 7			
7 + 2			
8 + 8			
9 + 10			
9 + 2			
2 + 6			
10 + 10			
2 + 8			

Addition Facts Inventory Record, Part 2

Lesson 5-9

DATE

Addition Fact	Know It	Don't Know It	How I Can Figure It Out
3 + 5			
3 + 6			
3 + 8			
3 + 9			
4 + 5			
4 + 7			
4 + 8			
4 + 9			
5 + 6			

Addition Facts Inventory Record, Part 2 (continued)

DATE

Addition Fact	Know It	Don't Know It	How I Can Figure It Out
5 + 7			
5 + 8			
5 + 9			
6 + 7			
6 + 8			
6 + 9			
7 + 8			
7 + 9			
8 + 9			

Notes

Shape Cards 1

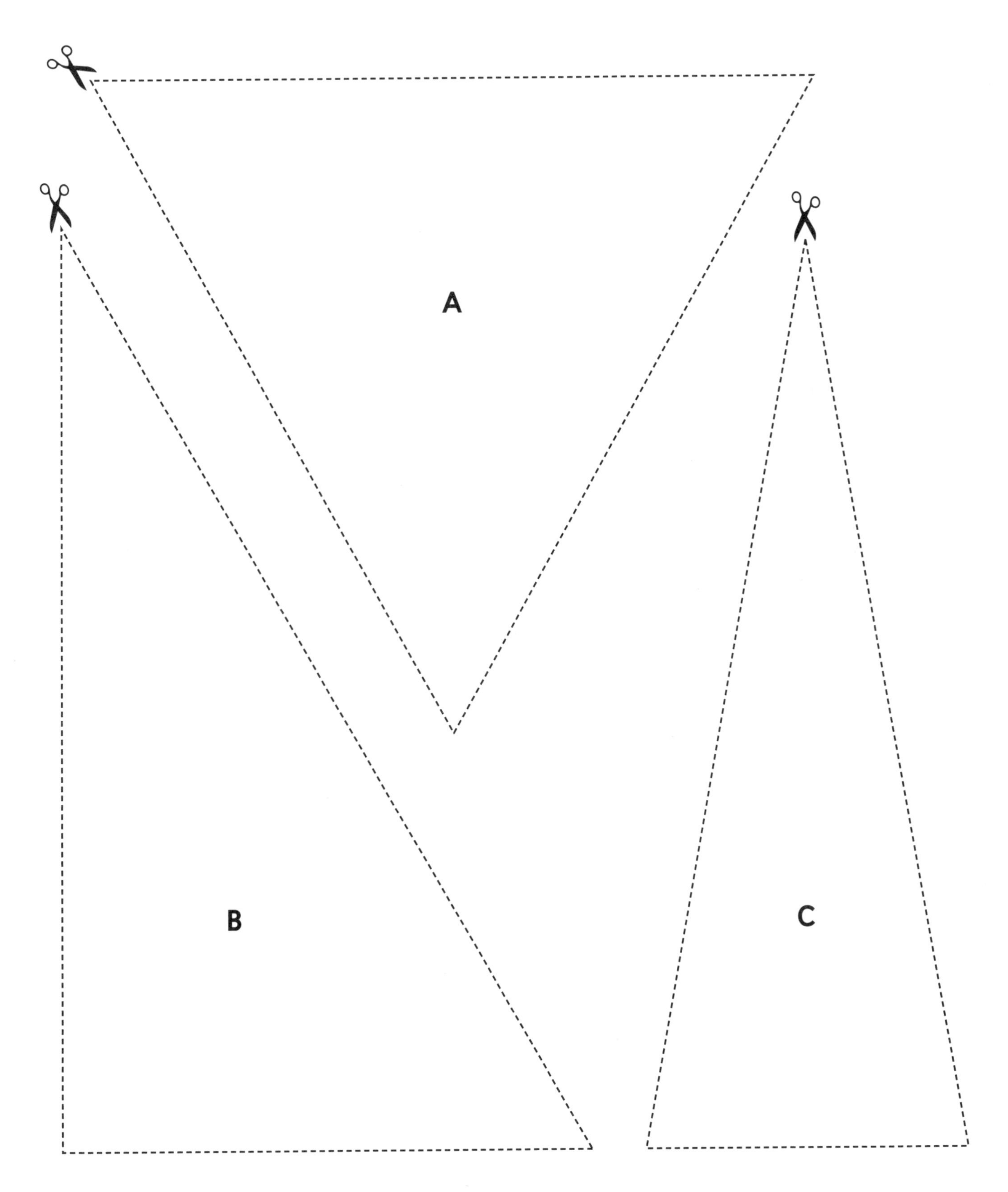

Shape Cards 2

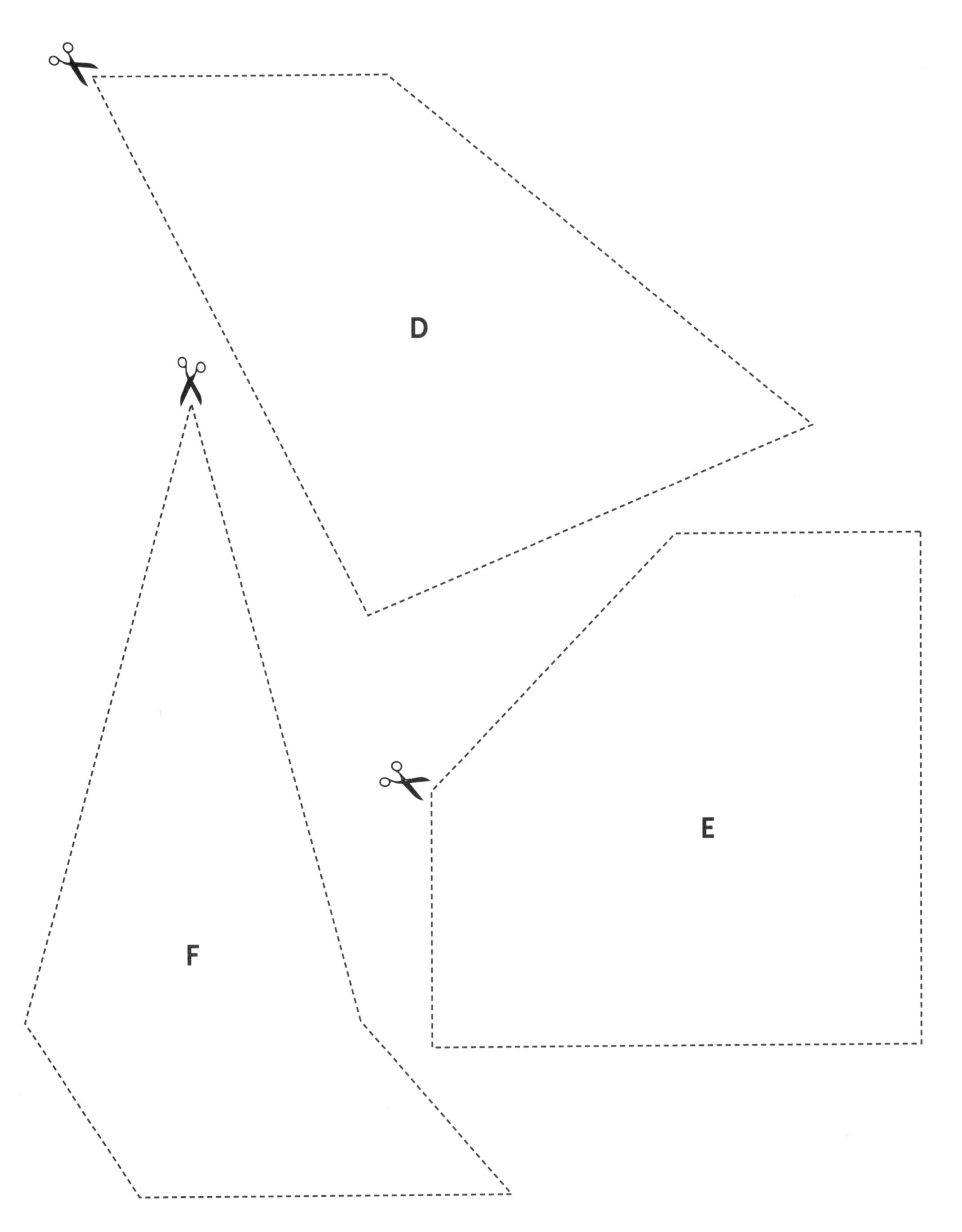

Shape Cards 3

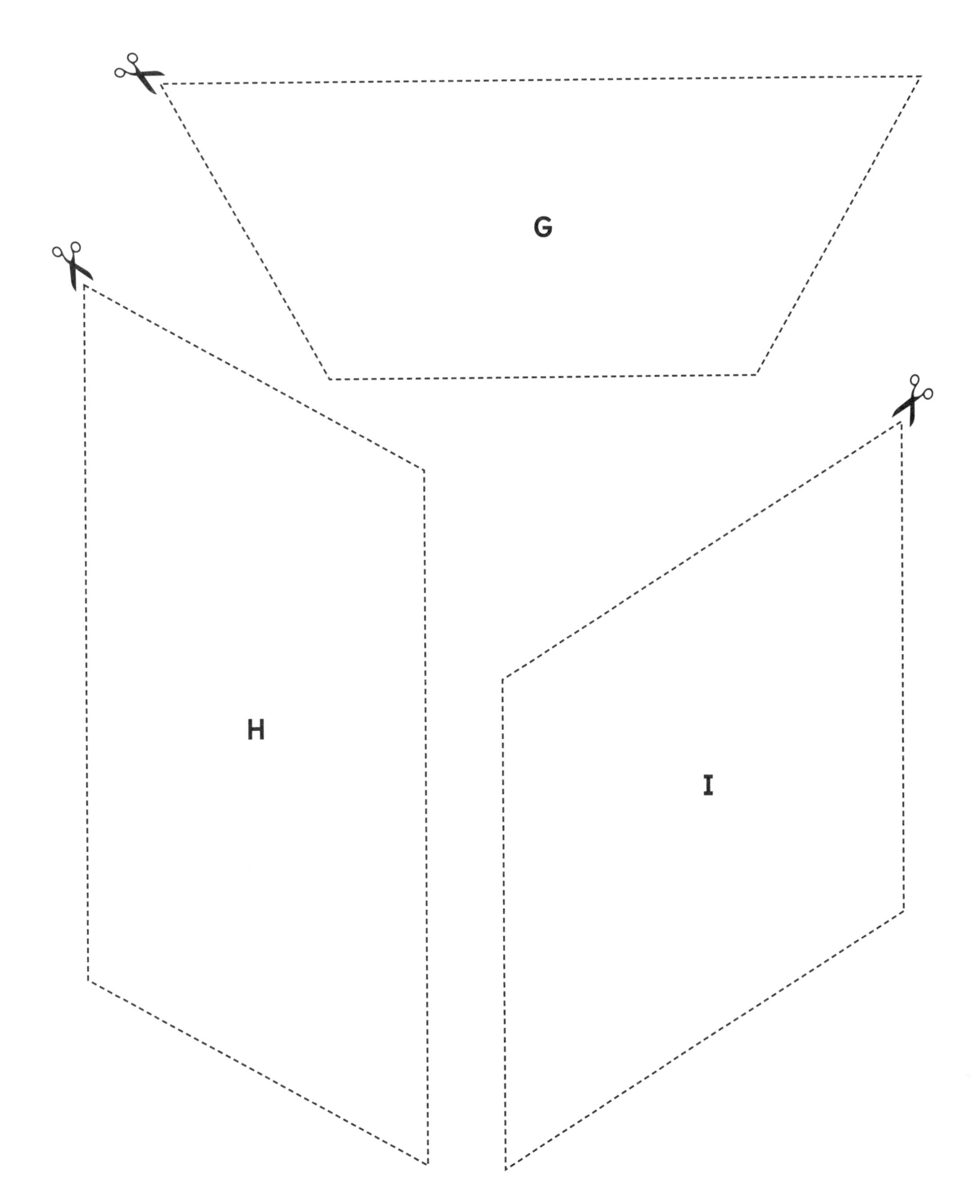

Shape Cards 4

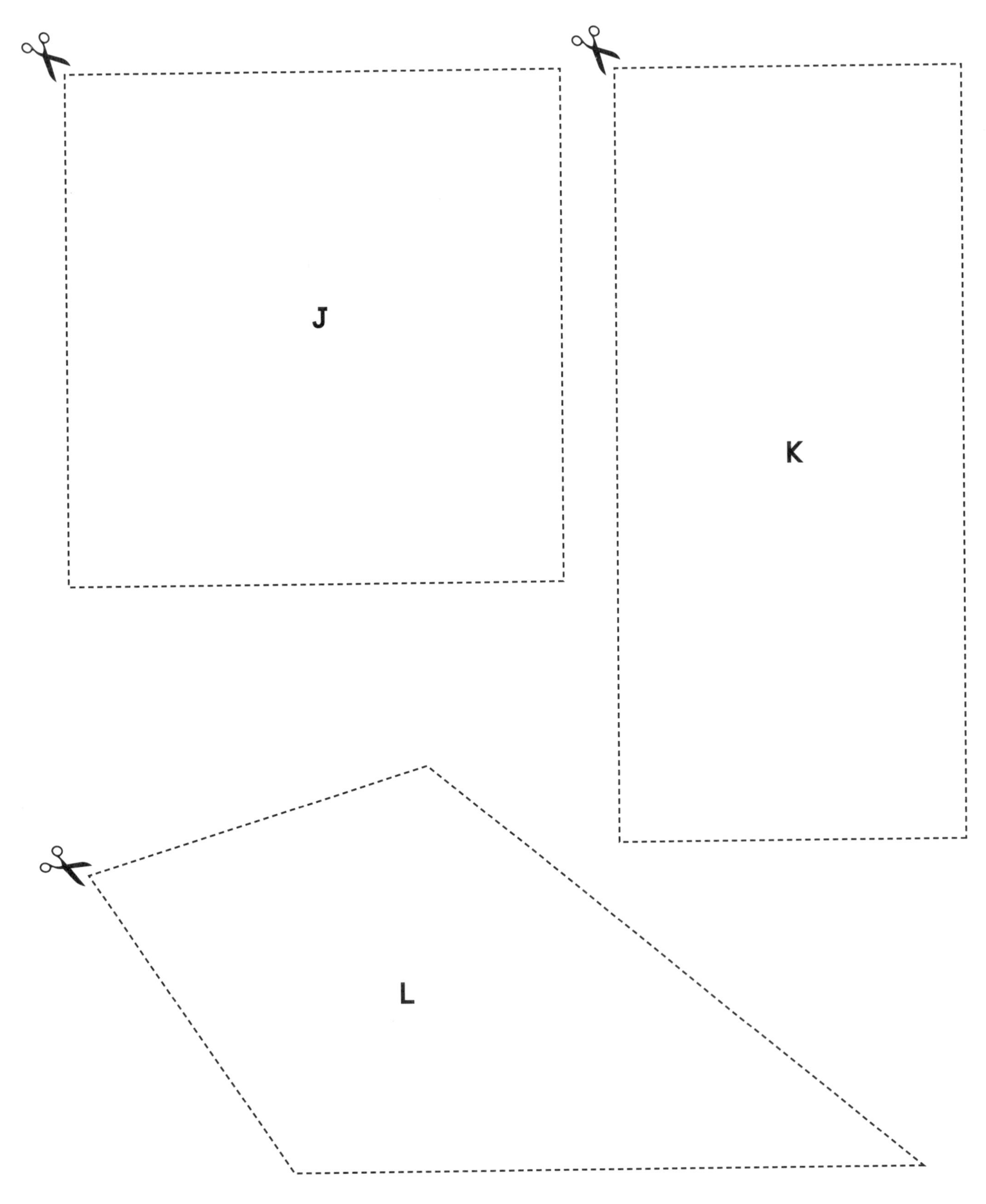

Shape Cards 5

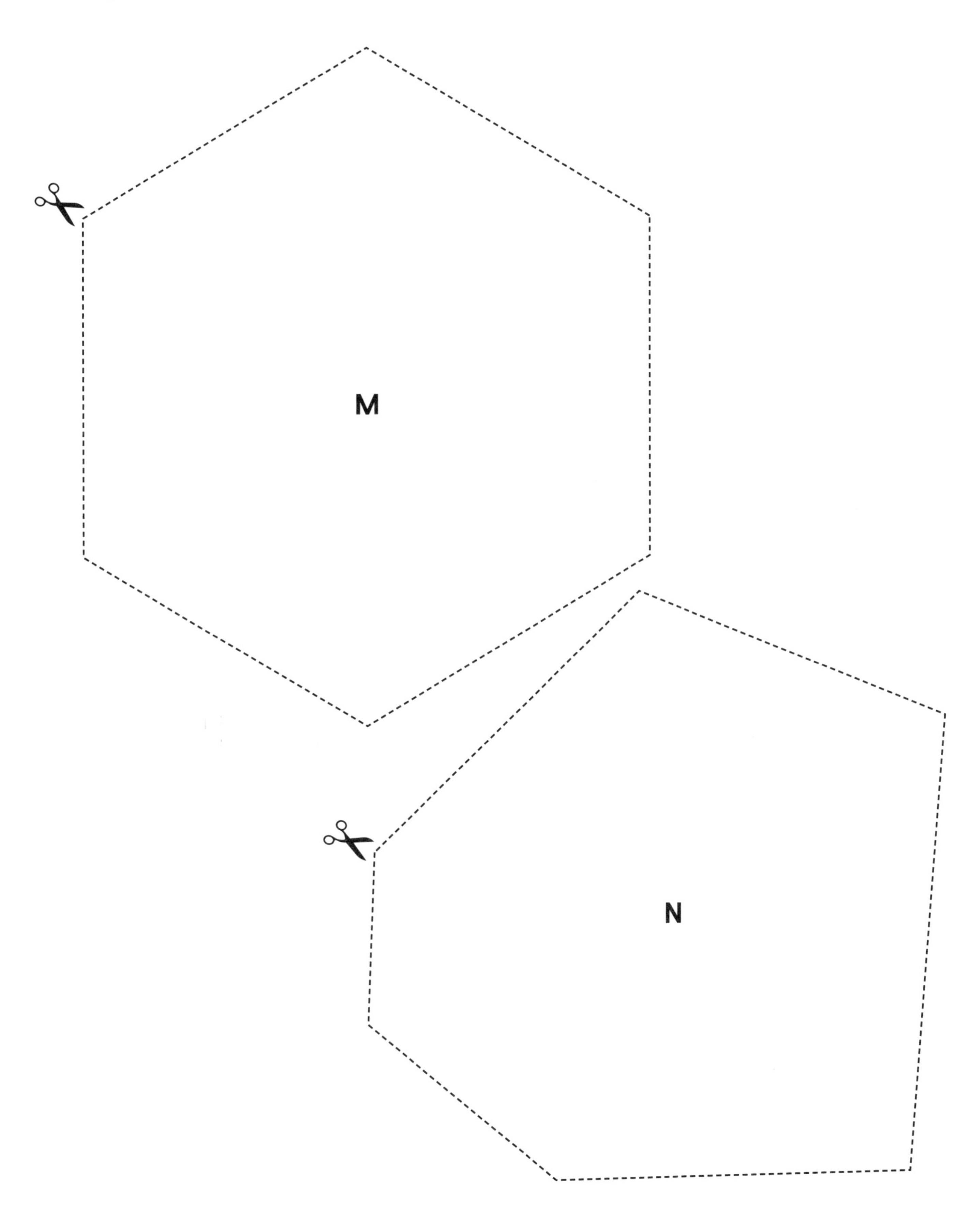

Shape Cards 6

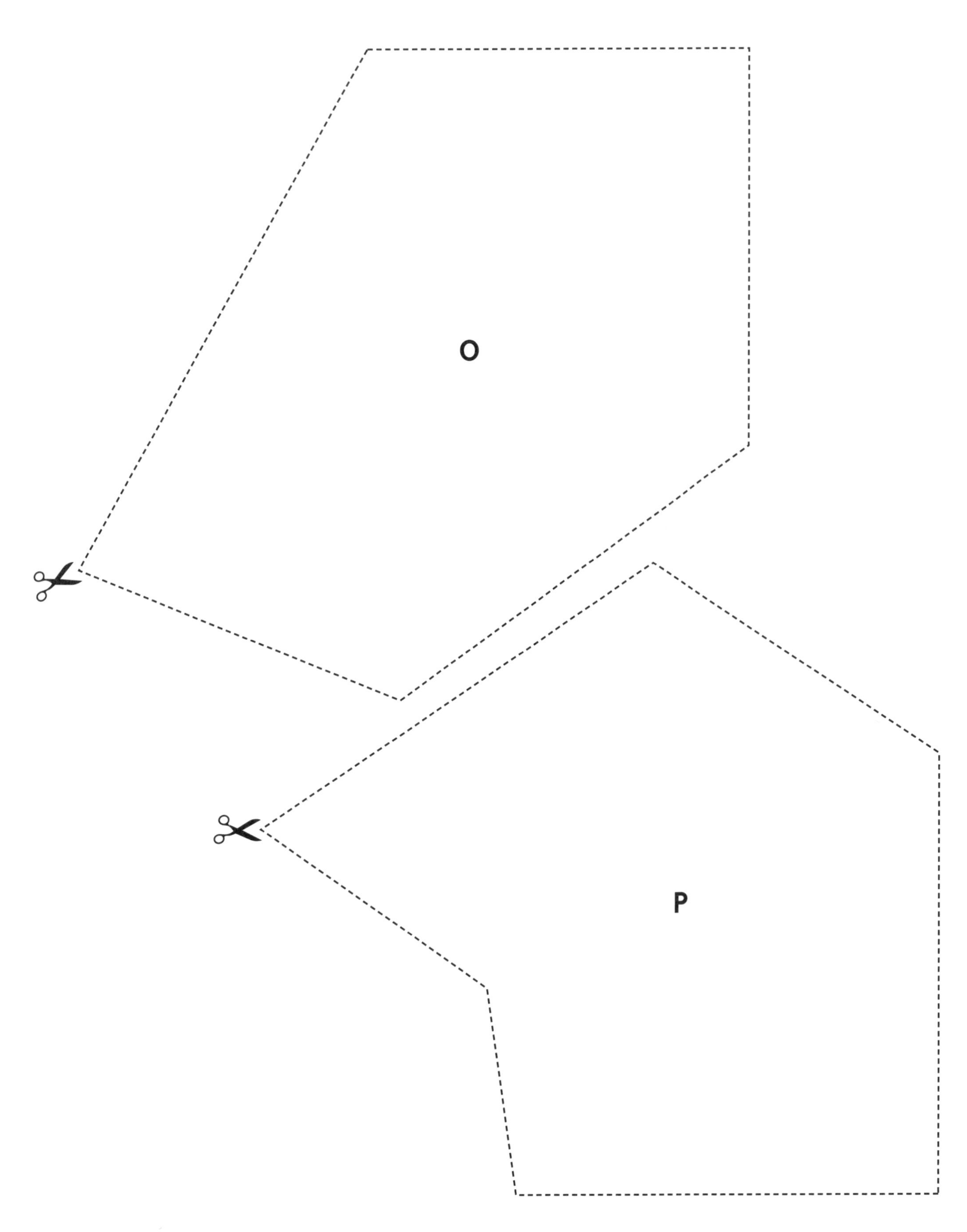

Shape Capture Attribute Cards

All of the angles are right angles.	There are no right angles.	There is only 1 right angle.
There are no parallel sides.	Only 1 pair of sides is parallel.	All opposite sides are parallel.
There are 3 sides, 3 vertices and 3 angles.	There are 4 sides, 4 vertices and 4 angles.	There are 5 sides, 5 vertices and 5 angles.
There are 6 sides, 6 vertices and 6 angles.	All sides are the same length.	Only some of the sides are the same length.